An Overview of FDA Regulated Products

From Drugs and Cosmetics to Food and Tobacco

An Overview of FDA Regulated Products

From Drugs and Cosmetics to Food and Tobacco

Edited by

Eunjoo Pacifici
International Center for Regulatory Science,
University of Southern California, Los Angeles, CA, United States

Susan Bain
International Center for Regulatory Science,
University of Southern California, Los Angeles, CA, United States

Academic Press is an imprint of Elsevier
125 London Wall, London EC2Y 5AS, United Kingdom
525 B Street, Suite 1650, San Diego, CA 92101, United States
50 Hampshire Street, 5th Floor, Cambridge, MA 02139, United States
The Boulevard, Langford Lane, Kidlington, Oxford OX5 1GB, United Kingdom

Library of Congress Cataloging-in-Publication Data
A catalog record for this book is available from the Library of Congress

British Library Cataloguing-in-Publication Data
A catalogue record for this book is available from the British Library

ISBN: 978-0-12-811155-0

For information on all Academic Press publications visit our website at https://www.elsevier.com/books-and-journals

Publisher: Mica Haley
Acquisition Editor: Erin Hill-Parks
Editorial Project Manager: Megan Ashdown
Production Project Manager: Mohanapriyan Rajendran
Designer: Greg Harris

Typeset by Thomson Digital

For
Dr. Alex Sevanian
Kyo Hyun and Hee Jae Kim
Choungil Kim

Contents

Contributors

Lilit Aladadyan
Tobacco Center of Regulatory Science (TCORS), University of Southern California, Los Angeles, CA, United States

Susan Bain
International Center for Regulatory Science, University of Southern California, Los Angeles, CA, United States

Paul Beninger
Tufts University School of Medicine, Boston, MA, United States

Roger Clemens
University of Southern California School of Pharmacy, International Center for Regulatory Science, Los Angeles, CA, United States

Mary Ellen Cosenza
MEC Regulatory & Toxicology Consulting, LLC, Moorpark, CA, United States

Daniela Drago
George Washington University, Washington, DC, United States

David A. Dzanis
Regulatory Discretion, Inc., Santa Clarita, CA, United States

Shayesteh Fürst-Ladani
SFL Regulatory Affairs & Scientific Communication, Basel, Switzerland

Nanae Hangai
Sanofi, Paris, France

Michael Jamieson
International Center for Regulatory Science, University of Southern California, Los Angeles, CA, United States

Christy Kadharmestan
College of Law, Michigan State University, East Lansing, MI, United States

Michael M. McGuffin
American Herbal Products Association, Silver Spring, MD, United States

Eunjoo Pacifici
International Center for Regulatory Science, University of Southern California, Los Angeles, CA, United States

Nancy Pire-Smerkanich
International Center for Regulatory Science, University of Southern California, Los Angeles, CA, United States

Frances J. Richmond
International Center for Regulatory Science, University of Southern California, Los Angeles, CA, United States

Jonathan M. Samet
Colorado School of Public Health, Aurora, CO, United States

Sai S. Tatavarty
Abbott Diabetes Care, Alameda, CA, United States

Simone E. Turnbull
Sanofi, Chattanooga, TN, United States

James William Woodlee
Kleinfeld, Kaplan and Becker LLP, Washington, DC, United States

Anthony L. Young
Kleinfeld, Kaplan and Becker LLP, Washington, DC, United States

Foreword

The products regulated by FDA constitute over 20% of all consumer-spending in the United States, amounting to several trillion dollars per year, with a notable impact on the global economy. Fundamental regulatory principles provide the basis for the Agency's role and its work to enhance public and individual health.

However, new and emerging life science discoveries, innovation in product development, advances in manufacturing and commerce, as well as socioeconomic changes represent rapidly evolving challenges and opportunities to regulatory science. Providing knowledge and offering dialogue, guidance, and instruction for the safe and effective development and use of FDA-regulated products reduces uncertainties for all stakeholders, including innovators and investors. This book addresses both the challenges and opportunities, by connecting information and reference material to the principles in FDA regulation with selected practical examples of regulatory case work, derived from key technology and fast-paced product development categories. Experienced educators and regulatory professionals have provided a remarkably comprehensive, user-friendly book for a broad audience. But more importantly, it also prepares for the future by guiding consumers, patients, and professionals in the food and health industries on how to interpret and apply information so that it leads to tangible results.

Readers will therefore be equipped with the practical knowledge to more effectively participate in critically important dialogues with the FDA, that will support the Agency's goal to make regulatory decisions based on the best available data, information, and knowledge. This investment in education and training in regulatory science will advance FDA from an Agency historically focused on policing and enforcing, to a science-driven, expert-guided authority that uses proactive innovation to keep up with technological change and communicates responsively and transparently with all interest groups.

Frank F. Weichold, MD, PhD
Director, Critical Path and Regulatory Science Initiatives
Office of Regulatory Science and Innovation (ORSI)
Office of the Chief Scientist (OCS)
Office of the Commissioner (OC)

Preface

Is toothpaste considered a cosmetic or a drug? What about sunscreens? Does the FDA approve food products before they can be sold? Are drugs that are approved for human use also considered safe and effective for use in animals? The answers to these questions are not apparent to most people. And for those interested in working in the food, health, and medical product industry, it is reassuring to know that there is a high demand for professionals trained in regulatory science who can connect the legal and regulatory requirements to scientific evidence.

Regulatory science as an academic discipline is relatively new. And it is by nature, interdisciplinary, where science, business, and law intersect. For many not familiar with the field, it can be quite daunting when trying to navigate the subject in a maze of laws, regulations, products, and companies. As educators, we have seen first-hand how simple explanations can help the students grasp fundamental concepts. It is in this spirit that we conceived of developing a book that would provide the novice reader with a practical guide to a wide array of FDA regulated products.

To tackle this project, we assembled authors who are experts in their fields. Most, in fact, lecture in our introductory level courses and understand the needs of our target audience as well as concepts that are most difficult for students to grasp. We, therefore, instructed our authors to consider approaching their writing as if they were lecturing to their students. Please note that the writings in this book represent the personal views of authors and not those of their employers.

There are other books that explore particular aspects of regulated products, including drug discovery and development, and clinical trial management. There are also books that cover legal and regulatory requirements for medical products and some that discuss business issues of the biopharmaceutical industry. However, currently there is no single book that provides a comprehensive overview of regulatory frameworks for a broad range of product categories. Our goal was to create a book that is both concise and easily accessible for a wide and diverse audience.

Today's challenge, especially for many newcomers to the regulated industry, is not necessarily to gather regulatory information, but to know how to interpret and apply it. The ability to discern what is important from what is not, and to interpret regulatory documents correctly, provides a valuable competitive advantage to any professional in this field. Our book aims to convey a summary of the key information that individuals with an interest in the regulated industry should know, and—more importantly—to unveil the meaning of critical regulatory concepts.

Although similarly structured, each chapter can also stand on its own to tackle a single product category. Hence the reader can enjoy reading the entire book from beginning to end or select an individual chapter to focus on a particular topic. Hopefully, you will find this field as fascinating and interesting as we do and will be inspired to further your studies.

Eunjoo Pacifici
Susan Bain

Acknowledgments

We are guided by our students, past, present, and future, from all corners of the world, who teach and inspire us every day with their curiosity, courage, and optimism; and their desire to contribute to the health and well-being of humans and animals.

This book would not have been conceived without a nudge from Frances Richmond, who thought it would be fun to develop an approachable book for this not-so-approachable subject. Frances has the most admirable talent for delicately nudging people to places that they never envisioned for themselves and yet upon their arrival realize that it was their destination all along. For that and so much more, we will forever hold her in our hearts with the utmost affection and gratitude.

We are indebted to our chapter authors who, with their passion for teaching and generosity of their hearts, agreed to embark on this journey with us. From the moment you gave your tentative "nod" to the submission of your final manuscript, you were gracious, responsive, and diligent despite all your competing obligations. Thank you, thank you, thank you!

A big "thank you" goes out to Daniela Drago, our coauthor, who helped with the original book proposal and the initial organization of the book. Our final product has benefited from your valuable input.

We thank Amelia Spinrad and Vaibhavi Chokshi for recommending quotes and graphics to include in the book, and for their assistance in reviewing drafts at various stages. We also thank Sahana Roopkumar for her hard work in creating the fictitious labels for veterinary products. Her creativity, hard work, and attention to detail is much appreciated.

It has been a pleasure to work with everyone at Elsevier. The project passed through many different hands but was managed seamlessly during the entire process. We give special thanks to Kristine Jones, who facilitated and guided the development of the original book proposal; and Megan Ashdown, our Editorial Project Manager, who has been resourceful, supportive, and helpful every step of the way.

We thank our children (Sarina, Noah, John, and Jennifer) for their great encouragement and the willingness to even review a few of the chapters. And finally, we each want to recognize the person whose love and support is weaved into the fabric of who we are and what we do. For Susan, that person is her father, Jim Waltrip, who continues to be her biggest fan and will be most delighted to receive his signed copy. For Eunjoo, it is of course her husband, Robert, from whom love was all that was asked but so much has been received.

Eunjoo Pacifici, Susan Bain
June 2018

CHAPTER

Introduction to FDA-regulated products

1

Eunjoo Pacifici, Susan Bain
International Center for Regulatory Science, University of Southern California, Los Angeles, CA, United States

When a company creates a product that directly or indirectly adversely impacts the health of people, that product must be regulated. The process by which it's created must be regulated.

– Kenneth C. Griffin

An Overview of FDA Regulated Products. http://dx.doi.org/10.1016/B978-0-12-811155-0.00001-6

CHAPTER OBJECTIVES

After reading this chapter, the reader will be able to:

- identify the market size represented by the products regulated by the United States Food and Drug Administration (US FDA),
- describe different categories of regulated products,
- recognize and compare how different products within a product category are regulated differently,
- identify therapeutic claims and which products can make them,
- describe the history of the modern regulatory system, and
- identify the key concepts that underlie the current regulatory framework.

Products regulated by the US FDA represent close to one-fifth of all products sold in the country or more than $2.4 trillion (United States Food and Drug Administration, 2017a). Globally, the pharmaceutical sector alone garners over a trillion dollars in sales. The public uses these products for nourishment, to maintain health and to derive therapeutic benefit. For the industry that develops and commercializes these products, it is important that their products address the needs of the public, are commercially viable, and meet the regulatory requirements of the FDA and, if they are sold abroad, of all the regulatory authorities of the relevant regions. With the continuing growth of this sector, the globalization of the industry, and the challenges of meeting regulatory requirements, the need for trained professionals in this field is increasing. As colleges and universities in the United States and elsewhere establish programs to prepare their students to enter the FDA-regulated industry, there is also an emerging need for a book that can be used in conjunction to aid in the instruction of the students. In addition, professionals working in other sectors like investment firms and media outlets may also be interested in the subject as they cover regulated industries and products. Therefore, this book was conceived to help interested novices gain a basic understanding of how products are regulated by the FDA.

We have focused primarily on the US system because, up to now, the United States represents the single largest market for many of the products and also because the US system is one of the oldest and most well-established regulatory systems. Other countries have their own regulatory authorities with their own requirements for gaining market authorization of regulated products. Although efforts to harmonize regulatory requirements have been ongoing for many decades, we are far from having universally harmonized standards. Levels of harmonization vary across countries and regions. For example, some countries require local clinical trials be conducted for a new product to understand its effects on local populations.

We have, in selected chapters, for example, "Drugs" and "Medical Devices", included other agencies for the purpose of comparisons. In addition, Chapter 2 provides an insight into the organizations and approaches used by three prominent agencies: US FDA, Japan's Pharmaceuticals and Medical Device Agency (PMDA), and the European Medicines Agency (EMA).

1.1 WHAT IS A REGULATED PRODUCT?

A product is considered regulated if a governmental authority determines when and how the product is allowed to be commercialized. For example, a regulatory authority may specify requirements for quality, safety, and effectiveness to be demonstrated for a product before granting marketing authorization and may have additional requirements to keep the product on the market.

1.2 HOW ARE DIFFERENT PRODUCTS REGULATED?

In the United States, the FDA separates products into specific categories: food, dietary supplements, cosmetics, drugs, biologics, medical devices, veterinary products, and tobacco. Food, dietary supplements, and cosmetics fall under the jurisdiction of Center for Food Safety and Nutrition (CFSAN), drugs under Center for Drug Evaluation and Research (CDER), biologics under Center for Biologics Evaluation and Research (CBER), medical devices under Center for Devices and Radiological Health, veterinary products under Center for Veterinary Medicine (CVM), and tobacco under Center for Tobacco Products (CTP).

The main law that governs these products in the United States is the Federal Food, Drug, and Cosmetic Act (FDCA), which was established in 1938 and amended as needed over the years to strengthen the regulatory framework (see Chapter 2). The laws are passed in the United States as Acts of Congress and organized, or codified, into United States Code (USC). Of the 53 titles in the USC, title 21 corresponds to the FDCA (Legal Information Institute, 2017). To operationalize the law for enforcement, governmental agencies like the FDA are authorized to create regulations. The Code of Federal Regulations (CFR) details how the law will be enforced. The CFR is divided into 50 titles according to the subject matter. Of those, title 21 corresponds to food and drugs. Therefore, professionals working in the FDA-regulated sector have three types of references for regulatory compliance: FDCA, 21USC, and 21CFR. As can be seen in Tables 1.1 and 1.2, there are different numbering schemes corresponding to each of the three references.

Generally, products within each category are regulated according to the risks they pose. For example, heart valves are more rigorously regulated than stethoscopes although both are categorized as medical devices. Hence, the two devices follow different regulatory paths for commercialization (see Chapter 5). Similarly, not all drugs are regulated the same way. A brand-new drug is an unknown entity and poses greater potential risks than a generic drug that is a copy of the original that has been on the market for many years. Thus, there are different regulatory requirements for different categories of drugs. Dietary supplements often look like drugs in that they come in the form of tablets, capsules, and liquids in packaging with labels that make them look like drugs. But in the United States, dietary supplements are intended to supplement the diet and hence classified under the "umbrella" of foods (United States Food and Drug Administration, 2015a). Dietary

Table 1.1 Description of 21CFR Parts

21CFR Parts	Content
1–99	General
100s	Food for human consumption
200s	Drugs: General
300s	Drugs for human use
400s	Reserved
500s	Animal drugs and feeds
600s	Biologics
700s	Cosmetics
800s	Medical devices
900s	Mammography quality standards
1000s	Radiological health
1100s	Tobacco products
1200s	Regulations under other acts administered by the FDA

Table 1.2 FDCA and 21USC Section Numbers for Selected Items

FDCA Section	21USC Section	Description
501	351	Adulterated drugs and devices
502	352	Misbranded drugs and devices
503	353	Exemptions and consideration for certain drugs, devices, and biological products
505	355	New drugs
506	356	Fast-track products
510	360	Registration of producers of drugs or devices
512	360b	New animal drugs
526	360bb	Designation of drugs for rare diseases or conditions

supplements do not require premarket approvals and are regulated by CFSAN. Tobacco products are the most recent addition to fall under the jurisdiction of FDA and are subject to a unique regulatory framework in that these products pose only risks and no benefits. Chapter 11 examines in detail how tobacco products are regulated by the FDA. Table 1.3 shows the different regulatory routes for medical products. As you read the corresponding chapters, keep the comparisons below in mind regarding the types of products and the risk-based oversight of products within each category.

Table 1.3 Regulatory Routes for Medical Products

	Medical Devices	Drugs	Biologics
Responsible FDA center	CDRH	CDER	CBER/ CDER
Regulatory route(s)	510K waived, 510K notification, PMA	OTC, ANDA, NDA	BLA
Clinical trial initiation	IDE	IND	IND

1.3 PRODUCT CLASSIFICATION

In the United States, the FDCA provides definitions for the different product categories along with allowable claims. For example, drugs, biologics, and medical devices can make therapeutic claims like "treatment of cancer" or "reduces symptoms associated with arthritis." Therapeutic claims also include implied statements like "relieves bronchospasm" or "relieves congestion." It is illegal for nonmedical products, like dietary supplements and cosmetics, to make therapeutic claims. If the FDA discovers that a company has made those statements, the FDA may respond by sending the company a warning letter with a notice to act within 15 working days to correct the violation. Structure/function claims are shared across food, dietary supplements, and drugs. Examples include "calcium builds strong bones" and "fiber maintains bowel regularity." If a product looks like a cosmetic but claims to treat eczema, it is a drug not a cosmetic and needs to meet all the regulatory requirements for a drug.

A product may be determined to be a drug if it contains a known drug ingredient. Hence, if an herbal remedy (a dietary supplement) contains sildenafil (the active ingredient in the prescription drug Viagra), it falls in the category of drugs and is not allowed to be marketed as a dietary supplement.

Even if a product does not contain any drug ingredients, its intended use may cause it to be categorized as a drug. A product's intended use may be determined through the company's printed materials like advertisements and promotional materials or statements made by company representatives as well as information posted on the product website. Hence, even if the dietary supplement mentioned above does not contain sildenafil, it would be categorized as a drug if it claims to improve male sexual dysfunction.

Drugs that have high potential for abuse with no accepted medical use are illegal and cannot be imported, manufactured, distributed, possessed, or used. In the United States, the "Controlled Substances Act" is the law that oversees these dangerous products and the Drug Enforcement Agency (DEA) is the federal agency that enforces the law (United States Drug Enforcement Administration, 2016).

Sometimes, it is not clear as to which class a product belongs. For example, what if a dental bone grafting material is combined with a bone growth factor? Is it a device or a biologic? In this case, the product would be classified as a combination product with CDRH as the lead center for regulating the product. Chapter 6 addresses these combination products and the concept of the "primary mode of action." One way to think about this question is to consider whether you would use one component without the other. The dental bone grafting material could be used without the bone growth factor in dental procedures to fill, augment, or reconstruct. But the growth factor alone would not be used for that intended use. What if a lipstick contains sunscreen? Is it a cosmetic or a drug? Today, many cosmetics like lotions and creams contain sunscreen components. In these cases, the products must be labeled as over-the-counter (OTC) drugs and meet OTC drug requirements. Chapter 3 examines the different classes of drugs and their regulatory pathways and Chapter 10 discusses cosmetics in more detail. Biologics include living cells and tissues as well as complex molecules that are, or similar to, natural substances such as enzymes, antibodies, or hormones. These products are different than traditional drugs, which are pure chemical substances with well-known structures manufactured through chemical synthesis. Although biologics are subject to federal regulation under the Public Health Service Act (PHS), they also meet the definition of "drugs" and are considered a subset of drugs. Hence, biologics are regulated under provisions of both PHS and FDCA (United States Food and Drug Administration, 2015b). Chapter 4 examines different types of biologics and how they are regulated.

1.4 HISTORY OF THE MODERN REGULATORY SYSTEM

Prior to the early 1900s, food and drugs were not regulated in the United States, and the marketplace saw widespread adulteration and misbranding of food and drugs. Today, the US regulatory system enforced by the US FDA is considered to be the most advanced around the world, with a staff of approximately 15,000 employees and an annual budget of $5.1 billion projected for 2018. The US FDA has broad jurisdiction that includes most food products (other than meat and poultry), human and animal drugs, biologics, medical devices, radiation-emitting products, cosmetics, animal feed, and tobacco products. Examining the evolution of the US regulatory system will provide the reader with a greater understanding of the key concepts that underlie the current framework (United States Food and Drug Administration, 2017b).

1.4.1 ADULTERATION AND MISBRANDING

Passage of major legislation in the area of food and medical products has occurred often following tragic incidents that revealed a need for, or further expansion of, government oversight. Prior to 1900, food products and drugs were routinely found to be substandard due to deceptive and fraudulent practices. For example, food products and drugs often contained low-cost substitutes or harmful additives (adulteration) without disclosing accurate information on the product labels (misbranding). Chemical preservatives and color additives were used without any control, milk was diluted with water, and quinine (drug to treat malaria) was mixed with ineffective ingredients. In Europe, Great Britain had already passed its national food and drugs act in 1860. The United States was slow to act but eventually passed its first landmark legislation, the Food and Drugs Act of 1906. This law prohibited the manufacture and interstate transportation of adulterated or misbranded food and drugs. Although the law was repealed and replaced by the Federal Food Drug and Cosmetic Act of 1938, the two terms, adulteration and misbranding, remain as key concepts in FDA regulation to this day. These terms, however, have broader implications in the modern era in that adulteration can include failure to follow good manufacturing practice (GMP) and misbranding can include misleading internet promotion and advertising. The original 1906 law also introduced the concept of conforming to the official drug standards as established by the United States Pharmacopeia (USP) and the National Formulary (NF). Advances in analytical chemistry during this era (1900s) were used to develop standards and assay methodologies.

But the 1906 law primarily focused on truth in labeling without requiring a product to demonstrate its safety or efficacy. There were no requirements for premarket submission of information, review, or approval. Hence, inferior and dangerous products remained on the market.

1.4.2 SAFETY

The next landmark food and drug legislation in the United States occurred after an elixir of sulfanilamide formulated in the toxic solvent diethylene glycol killed 107 people, many of whom were children. This incident revealed a major weakness in the existing system in that the product was legally marketed without a requirement to demonstrate its safety. The 1906 Food and Drugs Act was repealed and replaced with the Federal Food, Drug, and Cosmetic Act of 1938 (FDCA). The new law required that a new drug must provide scientific proof of its safety through a premarket review process, initiating the use of animal studies to establish product safety. This was the beginning of the New Drug Application (NDA) process. The law also prohibited the addition of poisonous substances to food products. In cases where such substances are unavoidable, safe tolerances were established. The law authorized factory inspections by the FDA and established food standards. In addition, cosmetics and medical devices were brought under the jurisdiction of FDA. Although the 1938 law strengthened the US regulatory system by introducing the concept of premarket review, it did not require new drugs to demonstrate efficacy.

1.4.3 EFFICACY

When reports surfaced that babies were being born in Europe with severe limb deformities due to the toxic effects of a popular tranquilizer, thalidomide, the United States was already considering further strengthening its regulatory system. The thalidomide incident had less impact in the United States because the product was still under evaluation by the FDA, but it helped to propel the passage of the Kefauver–Harris

Amendments of 1962. The new law amended the FDCA to require drug manufacturers demonstrate not only that a product is safe, but must also provide substantial evidence of effectiveness for its intended use. The amendments required the evidence be derived from adequate and well-controlled studies, a concept that provided a new and enduring standard, of two well-controlled studies, for conducting clinical trials. Other important elements included the requirement for premarket approval of a new drug and corresponding labeling, informed consent of study subjects, adverse event reporting to monitor safety of new drugs, and prescription drug advertisement to include side effect information. The amendments also allowed FDA to establish GMPs and to regularly inspect manufacturing facilities.

1.4.4 REGULATION OF MEDICAL DEVICES

Although medical devices were placed under the FDA's authority in 1938 as described above, the types of medical devices available at the time were technologically simple, and the regulatory framework for these products were not well defined. Like drugs, adulterated or misbranded products were outlawed, but there was no requirement for premarket testing, review, or approval. A comprehensive law in the form of the Medical Device Amendment of 1976 was passed after a study found that faulty medical devices (artificial heart valves, cardiac pacemakers, and intrauterine contraceptive devices) were responsible for 10,000 injuries including 731 deaths during a 10-year period. The amendment classified and regulated medical devices according to their potential risk. This risk-based classification of medical devices remains today and has been adopted globally, although the exact classification scheme may differ from that of the United States.

1.4.5 USER FEES

With the strengthening of the US regulatory system, heavier burden was placed on the FDA to regulate the products throughout their entire lifecycle. With each new law and regulation, the FDA's roles and responsibilities expanded. And without additional resources to accompany added responsibilities, the FDA could not make regulatory decisions in a timely manner. For example, by the early 1990s, the FDA reviews of new drug applications were taking 2 years or longer (United States Food and Drug Administration, 2010). The solution came in 1992 in the form of a new law—the Prescription Drug User Fee Act (PDUFA). The user fees are intended to provide the FDA with additional resources. In exchange, the FDA is required to meet performance goals, including timelines for reviewing product applications and publication of guidance documents. In addition, the funds from PDUFA allow for increased FDA–sponsor interactions throughout the drug development and review process. And because PDUFA needs to be renewed every 5 years, the regulatory system can be modified with each renewal. In fact, the performance focus of the program has expanded with every PDUFA renewal to date. The original PDUFA in 1992 only included the goals of eliminating backlog of new drug applications and completing

90% review target of priority applications in 6 months and standard application in 12 months. But PDUFA II in 1997 included additional performance goals related to industry meetings, special protocol evaluation, and electronic submissions. Subsequent renewals added provisions related to risk management, biosimilars, pediatric product development, and patient-focused drug development.

The user fee program, which started with drug reviews, has now expanded to include other product categories as listed in Table 1.4. Discussions are currently underway to institute a user fee program for the OTC monograph process as well. Other countries may have their own versions of user fee programs as regulatory authorities around the world struggle to secure resources to regulate novel products developed through advances in science and technology.

1.5 FOCUS ON POSTMARKETING OVERSIGHT

In recent years, the FDA has been expanding its focus into the postmarketing space and scrutinizing the safety and efficacy of products after they have been placed on the market. This move is to ensure that a product's benefit–risk profile remains acceptable throughout its lifecycle. Although the FDA requires demonstration of safety and efficacy through clinical testing prior to the approval of a new drug, the premarket studies may not reveal all the potential risks and further studies may be necessary to gain a better understanding of the safety of the product and efficacy profile. Furthermore, because premarket studies are performed in closely controlled settings, it is not clear whether the results seen in clinical trials would be replicated in real-world settings. In the United States, the FDA can require a company to conduct postmarketing research (PMR) in specific situations. For example, PMR may be required to verify clinical benefit for a product approved under the Accelerated Approval Pathway, which allows approvals based on surrogate or intermediate clinical endpoints. In addition, the FDA may require PMR to assess a known or suspected serious risk related to the use of the drug. With a growing number of companies utilizing expedited pathways to shorten development and review times of new drugs and devices, there is an increased reliance on data obtained from postmarketing research to determine, among other things, whether the benefits of a drug outweigh its risks. In addition, the

Table 1.4 FDA User Fee Programs (2018)

FDA User Fee Programs	Abbreviation	May Cost Up to (in $)
Prescription User Fee Act	PDUFA	2,421,495
Medical Device User Fee Amendments	MDUFA	310,764
Biosimilar User Fee Act	BsUFA	1,746,745
Generic Drug User Fee Amendments	GDUFA	1,590,792
Animal Drug User Fee Act	ADUFA	238,100
Animal Generic Drug User Fee Act	AGDUFA	193,000

FDA is working to expand its safety monitoring beyond spontaneous postmarketing reporting by developing a system through the "Sentinel Initiative" to obtain information from existing sources, including electronic health records, hospital records, insurance claims, and disease registries (Sentinel Coordinating Center, 2017).

1.6 THE MODERN REGULATORY SYSTEM

Recent trends in regulatory activities by the FDA point to a shift away from being reactive to being proactive. Although the major legislations during the FDA's first 100 years followed on the heels of high-profile incidents, more recent legislations have been spurred by calls for the agency to actively promote innovation, expand earlier access to therapies, and engage patients as key stakeholders. With the renewal of PDUFA occurring every 5 years, the United States now has a built-in mechanism to modify its regulatory framework to meet the needs of a changing world. The most recent renewal, PDUFA VI, also known as Food and Drug Administration Reauthorization Act of 2017 or FDARA, reauthorized PDUFA, MDUFA, GDUFA, and BsUFA (Sarata, Debrowska, Johnson, & Thaul, 2017). The law also contains provisions to encourage the development of drugs and devices for pediatric patients. For example, it allows the FDA to require a pediatric investigation into an adult cancer drug if the drug's mechanism of action is thought to be relevant to a pediatric cancer. Other provisions within the law include a regulatory path to allow OTC hearing aids, early consultation with the FDA on the use of new surrogate endpoints, and streamlining combination product reviews. The passage of FDARA follows another major law that was enacted at the end of 2016—"The 21st Century Cures Act" (Sarata, 2016; The 21st Century Cures Act). The "Title III—Development" of this law contains elements like patient-focused drug development, modern clinical trial designs, real-world evidence, patient access to therapies and information, and regenerative therapies that will have far-reaching impact on the industry and the patients.

REFERENCES

Legal Information Institute. (2017). *U.S. Code: Title 21—food and drugs*. Retrieved from https://www.law.cornell.edu/uscode/text/21.

Sarata, A. K. (2016). *The 21st Century Cures Act*. Congressional Research Service.

Sarata, A. K., Debrowska, A., Johnson, J. A., & Thaul, S. (2017). *FDA Reauthorization Act of 2017*. Congressional Research Service.

Sentinel Coordinating Center. (2017). Sentinel Initiative. Retrieved from https://www.sentinelinitiative.org/.

United States Food and Drug Administration. (May 3, 2010). *For industry*. Retrieved from https://www.fda.gov/ForIndustry/UserFees/PrescriptionDrugUserFee/ucm119253.htm.

United States Food and Drug Administration. (2015a). *Dietary supplement basics*. Retrieved from https://www.fda.gov/AboutFDA/Transparency/Basics/ucm193949.htm (06/08/2015).

United States Food and Drug Administration. (July 7, 2015b). *Frequently asked questions about therapeutic biological products*. Retrieved from https://www.fda.gov/drugs/developmentapprovalprocess/howdrugsaredevelopedandapproved/approvalapplications/therapeuticbiologicapplications/ucm113522.htm (07/07/2015).

United States Drug Enforcement Administration. (2016). Controlled Substances Act. Retrieved from https://www.dea.gov/druginfo/csa.shtml.

United States Food and Drug Administration. (December 21, 2017a). FDA basics. Retrieved from https://www.fda.gov/AboutFDA/Transparency/Basics/ucm553038.htm.

United States Food and Drug Administration. (October 3, 2017b). *About FDA*. Retrieved https://www.fda.gov/AboutFDA/WhatWeDo/History/Milestones/default.htm.

CHAPTER

2

Regulatory agencies of the ICH: Authorities, structures, and functions

Paul Beninger*, Nanae Hangai**

**Tufts University School of Medicine, Boston, MA, United States; **Sanofi, Paris, France*

Regulation represents a key means by which a government gives effect to its health policy preferences, especially through the exercise of a government's law-making powers.

– David Clarke (in World Health Organization's "Strategizing national health in the 21st century: a handbook")

An Overview of FDA Regulated Products. http://dx.doi.org/10.1016/B978-0-12-811155-0.00002-8

CHAPTER OBJECTIVES

After reading this chapter, the reader will be able to:

- describe the major historical events and key statutory responses that have shaped the development of the US FDA, Japan's PMDA, and the EU EMA,
- recognize and compare the essential differences in approaches that each agency takes in executing its responsibilities,
- recognize the important role that user fees play in supporting the timely review of applications for each jurisdiction, and
- describe the roles played by CIOMS and ICH in providing scientific and policy guidance in supporting the development of common approaches to regulatory processes.

2.1 INTRODUCTION

What are the regulatory agencies that are the public guardians of the medicinal and medical device products for human use? And how do they do their jobs? These are the key topics of this chapter. We will restrict our scope to the three major jurisdictions and agencies that were part of the original International Conference (now Council) on Harmonization (ICH, 2016) that was formed in 1990: the Japan's Pharmaceutical and Medical Device Agency (PMDA), the European Union European Medicines Agency (EMA), and the United States Food and Drug Administration (FDA). Information is taken primarily from the web sites of the respective agencies, focusing on the regulatory authorities, organizational structures, and key functions. Some of the topics will be examined in more detail in subsequent chapters that are specific to individual product classes and, hence, the information covered here will be at the top level.

2.2 UNITED STATES OF AMERICA: FOOD AND DRUG ADMINISTRATION

Although FDA celebrated its centennial in 2006 in recognition of the central role played by the Food & Drug Act of 1906, FDA's origins are actually rooted in the 1862 Bureau of Chemistry of the Department of Agriculture when it was charged with the investigation of adulterated foods (Janssen, 1981). Over the decades, FDA grew in an ad hoc way, as tragedies, catastrophes, and threats led Congress to assign, by statute, an increasing number of responsibilities to FDA regarding regulation of food safety, drugs, biologicals, the blood supply, vaccines, medical devices, and related products, so that today FDA regulates products that account for 25% of the $4 trillion consumer economy (FDA, 2016a, 2016b).

2.2.1 ESTABLISHING A REGULATORY MECHANISM

The issues that arose in the ensuing 150+ years were many times novel in that nowhere else in the world were there regulatory mechanisms already in place to serve as models for managing the types of problems that arose. So too, then, were the solutions novel, as FDA had to make organizational compromises with other agencies that shared responsibilities and subsequently had to work out feasible and workable regulatory mechanisms. For example, organizational compromises have included the complicated arrangements that exist for how FDA shares regulatory responsibility of certain foods with the Department of Agriculture, of water with the Environmental Protection Agency, of controlled substances with the Drug Enforcement Administration, and of advertising with the Federal Trade Commission (FDA, 2016c).

Regulatory mechanisms often broke new ground and established new paradigms: for example, safety, purity, and potency for biological manufacturing; misbranding and adulteration for product labeling; IND for clinical investigations and NDA for new drug applications; substantial evidence as a standard of evidence for approvability of an NDA; reporting requirements for the occurrence of adverse events; Rx and over-the-counter (OTC) categories of consumer access for new drugs; and risk-based regulation of medical devices (FDA, 2016d).

However, assignment of additional responsibilities was not matched with adequately allocated budgets that eventually led to a crisis of its own: gradually prolonged regulatory review times, lengthening to an average of 30 months during the 1980s. A paradigm shift came in the early 1990s with the establishment of user fees in a structured, 5-year renewal process that includes reviews of performance and establishment of new expectations. This schedule had the effect of moving from an almost exclusively reactive approach for making substantive changes to a more proactive approach for proposing new, important legislation. For example, the Patent Term Restoration component of the *Hatch–Waxman Act* of 1984 was enacted in response to a long growing crisis in prolonged review times described above and the concomitant loss of patent time protection (FDA, 2016e), whereas after the 5-year review had been established, the *FDA Modernization Act* of 1997 established financial incentives for conducting pediatric studies (FDA, 2016f); these were extended through the *Best Pharmaceuticals for Children Act* of 2002 (FDA, 2016g), renewed in 2007 and finally made permanent through *FDA Safety & Innovation Act* in 2012 (FDA, 2016h), after decades of failed efforts to convince industry to conduct needed studies with newly approved pharmaceuticals for children with unmet medical needs.

The globalization of products has put pressure on the globalizing processes themselves, specifically regulatory processes, including compliance. This development has been significantly helped by the work of CIOMS and ICH, discussed at the end of this chapter.

2.2.2 WHAT IS FDA'S MISSION?

FDA is responsible for protecting the public health by assuring the safety, efficacy, and security of human and veterinary drugs, biological products, medical devices, our nation's food supply, cosmetics, and products that emit radiation (FDA, 2016i). FDA is also responsible for advancing the public health by helping to speed innovations that make medicines more effective, safer, and more affordable and by helping the public get the accurate, science-based information they need to use medicines and foods to maintain and improve their health. Further, FDA has responsibility for regulating the manufacturing, marketing, and distribution of tobacco products to protect the public health and to reduce tobacco use by minors. FDA's jurisdiction includes the 50 United States, the District of Columbia, Puerto Rico, Guam, the Virgin Islands, American Samoa, and other US territories and possessions (FDA, 2016j).

2.2.3 HOW IS THE FDA STRUCTURED?

The FDA is an agency within the US Department of Health and Human Services. Other closely collaborating agencies in DHHS (2016) include the National Institutes of Health (NIH), the Center for Disease Control and Prevention (CDC), and the Center for Medicare and Medicaid Services (CMS). FDA is headed by the Commissioner, Food and Drugs. The Commissioner is one of approximately 1400 PAS employees in the federal government, that is, a *P*residential *A*ppointee, who requires *S*enate confirmation. The Office of the Commissioner includes the following direct reports: Chief of Staff, Chief Counsel, Chief Scientist, and Principle Deputy Commissioner (FDA, 2016k, 2016l). The FDA is organized under four super-Offices that oversee the core functions of the agency:

1. medical products and tobacco;
2. foods and veterinary medicine;
3. global regulatory operations and policy;
4. operations.

2.2.4 MEDICAL PRODUCTS

Medical products regulated by FDA can be classified into four basic categories: drugs, biologics, medical devices, and electronic products that give off radiation (FDA, 2016b).

Drugs include prescription drugs and some biologicals, including brand name, generic, and bio-similar products, nonprescription drugs, and OTC drugs. However, some aspects related to drugs are regulated by other agencies. For example, direct-to-consumer advertising is regulated by the Federal Trade Commission, and the Department of Justice's Drug Enforcement Administration (DEA) enforces the controlled substances laws and regulations of the United States, particularly with regard to the manufacture, distribution, and dispensing of legally produced controlled substances.

Biologics include vaccines for human use, blood and blood products like clotting factors, cellular and gene therapy products, tissue and tissue products, and allergens.

Medical devices include simple instruments like stethoscopes and scalpels; complex technologies, such as heart–lung machines and pace-makers; dental devices; plastic surgical devices, including breast implants; orthopedic implants and prosthetic devices; and in vitro diagnostic test kits, including companion diagnostics.

Electronic products that give off radiation include microwave ovens, X-ray equipment, including CAT scans, MRI machines, PET scans, laser products, including laser pointers, ultrasonic therapy equipment, including portable equipment, and sunlamps.

2.2.5 LEGAL FRAMEWORK

The Unites States Code (U.S.C.), which includes all of the currently enacted federal statutes, is maintained by the Office of the Law Revision Counsel in the US House of Representatives. Title 21 of the U.S.C. (FDA, 2016m) is reserved for FDA (21 USC FD&CA Chapter V: Drugs and Devices). There are 27 chapters: Chapter 9 is entitled *Federal Food, Drug, and Cosmetic Act* (*FFD&C Act*), enacted in 1938, as amended, which we are still under today (FDA, 2016n). *FFD&C Act* itself contains 10 chapters, of which Chapter 5 is Drugs and Devices. Important, selected Acts and amendments regarding drugs, biologics, and medical devices are shown in Table 2.1.

Biologics are additionally regulated under the *Public Health Service Act* of 1944, as amended, in 42 U.S.C. §262 (FDA, 2016o), 42 U.S.C. §351(a): Biologics Licensing Application (FDA, 2016p), and 42 U.S.C. §351(k) (biosimilar or interchangeable) (FDA, 2016q).

The Code of Federal Regulations (CFR) provides all of the regulations that implement the statutes enacted by Congress. Title 21 (FDA, 2016r) is reserved for FDA. Promulgation of regulations follows a process stipulated by the *Administrative Procedures Act* of 1946 that requires publication of a Notice of Proposed Rulemaking in the Federal Register (Bachorik, 2016). FDA may provide 6 months or longer for stakeholders to comment on the proposed regulations. FDA then publishes a Final Rule in the Federal Register that includes a summary of the submitted comments and the Agency's response to revise or retain the proposed regulation. At this point, the published regulations have the full force and effect of the enacting legislation. As a practical matter, this means that FDA has the flexibility to exercise enforcement discretion when circumstances warrant, whereas FDA does not have such flexibility with statutory-specified requirements.

2.2.6 FDA'S FUNCTIONS

The primary regulatory activities that are carried out for each of the three sets of products for human use, that is, drugs, biologics, and medical devices, include the following:

Table 2.1 Important US Legislation Regarding Drugs, Biologics, and Medical Devices

Year	Amendment or Act	Importance
1902	Biologics Control Act	Creates Establishment License Application (ELA) and Product License Application (PLA) (which combine to become Biologic License Application (BLA) in 1998) and develops standards for safety, purity, and potency
1906	Pure Food and Drug Act	Establishes standards for adulteration and misbranding
1938	Food, Drug & Cosmetic Act	Creates New Drug Application (NDA) and Investigational New Drug (IND) application
1951	Durham–Humphrey Act	Creates the categories of prescription (Rx) and OTC new drugs
1962	Drug Amendments (aka Kefauver–Harris Amendments)	Establishes an efficacy requirement and a new standard of evidence for approval and requires adverse event reporting, fair balance in advertising, and informed consent for clinical trials
1976	Medical Devices Amendments	Creates risk-based classification for review of medical devices
1983	Orphan Drug Act	Establishes criteria for orphan drugs
1984	Drug Price Competition and Patent Term Restoration (aka Hatch–Waxman Act)	Establishes the basis for the generic drug industry and allows recovery of development and review time for patents up to 5 years by originators
1992	Prescription Drug User Fee Act (PDUFA)	Establishes user fees for review of New Drug Applications, with reauthorization every 5 years
1997	FDA Modernization Act (FDAMA)	Enhances FDA's mission by increasing technological, trade, and public health features
2002	Medical Device User Fee and Modernization Act	Establishes user fees for review of medical devices
2002	Best Pharmaceuticals for Children Act (BPCA)	Establishes mechanisms for studying on-patent and off-patent drugs in children by offering financial incentives
2003	Pediatric Research Equity Act (PREA)	Establishes criteria by which companies must conduct pediatric studies
2007	FDA Amendments Act (FDAAA)	Reauthorizes PDUFA, MDUFA, BPCA, PREA
2012	FDA Safety and Innovation Act (FDASIA) Generic Drug User Fee Amendment	Reauthorizes user fees and created new criteria for expedited development and review of new drugs
2013	Drug Quality and Security Act (DQSA)	Establishes new compounding standards (Title I); establishes electronic, interoperable system to identify and trace prescription drugs as they are distributed in the United States (Title II)
2016	21st Century Cures Act	Introduces a range of provisions to enhance development of drugs, biologics, and medical devices
2017	FDA Reauthorization Act (FDARA)	Reauthorizes PDUFA for the 5th time, MDUFA for the 3rd time, and authorizes the Generic Drug User Fee Amendments (GDUFA) and the Biosimilar User Fee Act (BsUFA) for the first time

1. Permit unapproved products to undergo testing in humans (discussed below);
2. Review applications for marketing (discussed below);
3. Review labeling, advertising, and promotional materials;
4. Collect and evaluate adverse event information associated with the use of these products;
5. Audit operational activities, databases, and manufacturing sites for quality and compliance with regulations.

2.2.6.1 Permitting unapproved products to undergo testing in humans

The investigational new drug (IND) regulations, under 21 CFR § 312 (FDA, 2016s), authorize FDA to exempt companies from the statutory requirement that only approved drugs may be shipped in interstate commerce when the purpose of the shipment is to conduct a human study with an investigational drug. This means, in practice, that after a satisfactory review of an IND by FDA professional staff, a company is given permission to send properly manufactured and appropriately formulated product to any clinical investigators under FDA jurisdiction in order to conduct human studies according to an approved protocol. This usually includes provisions to determine patient eligibility, essential information for informed consent and institutional review, and an Investigator's Brochure, and may include additional provisions for a monitoring committee, an endpoint committee, or other supporting governance structures, depending upon the type of study (FDA, 2016t). The IND regulations under 21 CFR § 312 are equally applicable to biologics.

Because medical devices are classified according to risk, only devices that are determined to be of significant risk are required to undergo testing in humans, which requires an investigational device exemption (IDE), under 21 CFR § 812 (FDA, 2016u). The rationale for an "exemption" in an IDE is the same as for the IND described above for drugs and biologics.

2.2.6.2 Reviewing applications for marketing

The FDA is responsible for reviewing and approving new products for commercialization in the United States according to an established risk-based regulatory framework.

2.2.6.2.1 Drugs

The statutory pathways to market for drugs and generic drugs are shown in Table 2.2.

Key components of the New Drug Application (NDA) regulations, described under 21 CFR § 314 (FDA, 2016v), include the NDA process for new or modified new drugs, abbreviated applications for generic drugs, accelerated approval of new drugs for serious or life-threatening illnesses, and approval of new drugs in cases where human efficacy studies are not ethical or feasible.

There are four expedited programs in FDA for serious or life-threatening illnesses that can facilitate development and approval of drugs and biologics (FDA, 2016w):

1. Priority review: A designation under PDUF Act (1992).
2. Fast track: A designation under FD&C Act §506(b).
3. Breakthrough therapy: A designation under FD&C Act §506(a).

Table 2.2 Statutory Pathways to Market

Product	Statute	Section
Drugs		
New drug application (NDA)	US FD&CA	505(b)(1) 505(b)(2)
Abbreviated NDA (ANDA) [generics]		505(j)
Biologics		
Biologics license application (BLA)	US Public Health Service Act	351(a)
Biosimilar and interchangeable		351(k)
Medical Devices		
Class I general control (GC) requirements	US FD&CA	501, 502
Class II GC + 510(k) premarket notification		510(k)
Class III GC + premarket approval		513, 515

4. Accelerated approval: An approval pathway under FD&C Act §506(c).

   These are described in more detail in Chapter 3.

2.2.6.2.2 Biologics

The statutory pathways to market for biologics and biosimilars are also shown in Table 2.2 (FDA, 2016x). There are some regulations pertaining to biologics products found under 21 CFR § 600 (FDA, 2016y), including manufacturing establishment standards and inspections as well as reporting of adverse experiences. 21 CFR § 601 (FDA, 2016z) include licensing regulations for biologics and diagnostic radiopharmaceuticals. Similar to drugs as described above, there are provisions for accelerated approval of biological products for serious or life-threatening illnesses and approval of biological products when human efficacy studies are not ethical or feasible.

Additionally, regulations pertaining to product quality standards specific to biologics are provided under 21 CFR § 610 (FDA, 2016aa). They include release requirements, standard preparations and limits of potency, test for mycoplasma, testing requirements for communicable disease agents, dating period limitations, and labeling standards.

2.2.6.2.3 Medical devices

The statutory pathways to market for medical devices are shown in Table 2.2 (FDA, 2016bb). Three pathways to market are available to manufacturers of medical devices, depending on the risk classification of the device. Details can be found in Chapter 5 but briefly, the three paths to market for medical devices are as follows:

1. *Class I*: These are low-risk devices that are subject to general controls and which are applicable to all classes of devices. Example includes stethoscopes and scalpels.
2. *Class II*: These are moderate-risk devices for which general controls, by themselves, are insufficient to provide reasonable assurance of safety and

effectiveness. Products in this class require special controls to provide such assurance. Examples include orthopedic implants and cardiovascular catheters.

3. *Class III*: These are high-risk devices that typically require collection of clinical data under an investigational device exemption (IDE) and subsequent submission of a premarket application (PMA) prior to approval. Examples include heart valves and in vitro diagnostic kits that serve as companion diagnostics.

2.2.6.3 Pharmacovigilance

The thalidomide tragedy of the late 1950s—early 1960s was a watershed event for pharmacovigilance. The Kefauver–Harris Amendments that were passed by Congress included provisions for reporting adverse events by pharmaceutical manufacturers (FDA, 2016cc). Subsequent legislation over the decades extended activities into the development of case management, signal management and, most recently, into risk management (now appreciated as benefit–risk management). The FDA Amendments Act of 2002 authorized FDA to request a manufacturer to develop a risk evaluation and mitigation strategy (REMS) to ensure that the benefits of certain prescription drugs outweigh the risks (FDA, 2016dd).

2.2.7 OTHER PRODUCTS

The three categories of medical products for human use described above, drugs, biologics, and medical devices, are primarily managed by three different centers: Center of Drug Evaluation and Research (CDER), Center for Biologics Evaluation and Research (CBER), and Center for Devices and Radiological Health (CDRH) which are included in the Commissioner's Office as a super Office of Medical Products and Tobacco. These centers perform the same general set of activities although the organizational structures that support these functions may differ. There are many other products, outside of medical products for human use, which are regulated by the FDA. These include veterinary products, food, cosmetics, and tobacco products. Although these topics will be described individually in subsequent chapter, we provide a brief overview below.

2.2.7.1 Veterinary products

The Commissioner's Office includes the super Office of Food and Veterinary Medicine which includes the Center for Veterinary Medicine (CVM) (FDA, 2016ff). CVM provides Offices for the following major areas related to veterinary products: new animal drug evaluation; generic animal drug evaluation; surveillance and compliance; research; minor use, minor species animal drug development. Important statutes regarding veterinary products are provided in Table 2.3.

CVM is responsible for regulating livestock feeds, pet foods, veterinary drugs, biologics, vaccines, and devices. However, the US Department of Agriculture's Animal and Plant Health Inspection Service (APHIS) and Center for Veterinary Biologics regulate aspects of veterinary vaccines and other types of veterinary biologics.

Table 2.3 Important Legislation Regarding Veterinary Products

Year	Amendment or Act	Importance
1938	Federal Food Drug & Cosmetic Act	Gives authority to FDA to regulate animal drugs, biologics, and food additives when used in drinking water of food-producing animals
1959	Food Additive Amendments	Expands authority over additives to animal foods and drug residues in animal-producing foods
1968	Animal Drug Amendments	Introduces provisions to assure that animal drugs are safe and effective
1988	Generic Animal Drug and Patent Term Restoration Act	Requires manufacturers to submit abbreviated new animal drug applications for review and approval before marketing
1994	Animal Medicinal Drug Use Clarification Act	Permits veterinarians to prescribe extra-label uses of selected prior approved animal drugs and human drugs for use in animals under specified conditions
2003	Animal Drug User Fee Act	Establishes user fees for review of new animal drug applications
2004	Minor Use Minor Species Animal Health Act	Permits veterinarian use of medications to treat minor animal species and uncommon diseases of major animal species
2008	Animal Generic Drug User Fee Act	Establishes user fees for review of animal generic drug applications

2.2.7.2 Food

The super Office of Food and Veterinary Medicine also includes the Center for Food Safety and Nutrition (CFSAN), which provides the organizational structure for regulating food safety. CFSAN includes an Office structure for the following major areas related to food safety: food safety, food additive safety, nutrition and food labeling, dietary supplements, compliance; regulatory science, applied research and safety assessment, analytics and outreach. Important statutes regarding food safety are outlined in Table 2.4.

CFSAN also oversees infant formulas, bottled water, food additives, and dietary supplements. However, certain food products are regulated by the Department of Agriculture, such as some nongame meat and poultry, and egg products. And the Environmental Protection Agency (EPA) regulates aspects of drinking water. EPA develops national standards for drinking water from municipal water supplies (tap water) to limit the levels of impurities. In addition, the Department of the Treasury's Alcohol and Tobacco Tax and Trade Bureau (ATTB) regulates aspects of alcohol production, importation, wholesale distribution, labeling, and advertising.

Table 2.4 Important Legislation Regarding Food Safety

Year	Amendment or Act	Importance
1906	Pure Food and Drug Act	Bans interstate commerce in adulterated or misbranded foods
1938	Federal Food Drug & Cosmetic Act	Gives authority to FDA for the regulation of food safety
1958	Food Additives Amendment	Establishes category of "generally regarded as safe" (GRAS) for chemicals long-used as food additives that would not require further testing. New chemicals would require testing and the "Delaney clause" stipulates that chemicals found to cause cancer in humans or animals may not be used as food additive
1990	Nutrition Labeling and Education Act	Requires detailed information on food packaging as well as use of standardized descriptive terms, such as "low fat"
1994	Dietary Supplement Health and Education Act	Restricts FDA regulation of dietary supplements: Evidence of safety and efficacy are no longer required
2010	Food Safety Modernization Act	Modernizes food safety standards through emphasis on prevention, import safety, inspections, compliance and response, and enhanced inter-agency partnerships

2.2.7.3 Cosmetics

CFSAN includes the Office of Cosmetics and Colors. Two statutes important to the regulation of cosmetics include the *FD&C Act* of 1938 that gives authority to FDA for the regulation of cosmetics and defines cosmetics by intended use; and the *Fair Packaging and Labeling Act* of 1966 that sets out parameters for labeling cosmetic products (FDA, 2016gg). The following product areas are covered: color additives that are used in makeup and other products of personal care, cleansing agents of the skin, moisturizers, nail polish, and perfumes.

2.2.7.4 Tobacco products

The super Office of Medical Products and Tobacco includes the Center for Tobacco, which is responsible for activities related to tobacco products: regulations, science, health communication and education, and compliance and enforcement (FDA, 2016ii). The Family Smoking Prevention and Tobacco Control Act of 2009 gives FDA broad authority to regulate the manufacturing, distribution, and marketing of tobacco. The following tobacco products are covered: cigarettes, cigarette tobacco, roll-your-own tobacco, smokeless tobacco, and e-cigarettes.

2.3 JAPAN: PHARMACEUTICALS AND MEDICAL DEVICES AGENCY (PMDA)

Japanese regulation of pharmaceutical products began in 1980 as the *Relief Fund for Patients Suffering from Adverse Drug Reactions* (PMDA, 2016a). Multiple expansions of responsibilities and reorganizations resulted in the *Pharmaceuticals and Medical Devices Agency Act* (PMDA, 2016b) that began its activities on April 1, 2004 as can be seen in Table 2.5.

2.3.1 MISSION

PMDA has established a three-pillar system for the protection of the Japanese public. Each is institutionalized in an Office with an Executive Director (PMDA, 2016c).

2.3.1.1 Reviews

PMDA offers guidance and conducts reviews on the quality, efficacy, and safety of new drugs and medical devices through a system that integrates the entire process from preclinical research through approval, with the goal of reducing risk.

Table 2.5 Important Legislation Regarding Drugs, Biologics, and Medical Devices

Year	Legislation or Act	Importance
1979	Relief Fund for Patients Suffering from Adverse Drug Reactions	Created fund in response to patients who developed a unique adverse drug reaction syndrome, SMON (subacute myelo-optic neuropathy), following treatment with *Clioquinol* (iodochlorhydroxyquin)
1997	Pharmaceuticals and Medical Devices Evaluation Center of the National Institute of Health Sciences	Established product review activities
2004	Pharmaceuticals and Medical Devices Agency Act	Established pharmaceuticals and medical devices agency (PMDA)
2004	Pharmaceutical Affairs Law	Established marketing certification system in which controlled medical devices and in vitro diagnostics are certified by certification bodies registered by MHLW
2014	Advanced Health and Medicine Strategy Act	Encouraged regulatory science, which includes fostering collaborative relationships with academic institutions
2014	Securing Quality, Efficacy and Security of Pharmaceuticals, Medical Devices, Regenerative and Cellular Therapy Products, Gene Therapy Products, and Cosmetics Act	Established the reviewing standards for the primary products for treatment

2.3.1.2 Safety measures

PMDA collects, analyzes, and provides postmarketing safety information, with the goal of continually mitigating risk.

2.3.1.3 Relief services for patients suffering from adverse health effects

PMDA works to improve the public health in Japan by providing prompt relief to people who have suffered ill health effects resulting from adverse drug reactions or infections from biological products.

2.3.2 SCOPE

PMDA is responsible for pharmaceutical products, medical devices, in vitro diagnostics, and cosmetics. The Ministry of Agriculture, Forestry, and Fisheries is responsible for food (Food Safety Policy Division) and veterinary drugs (Animal Health Division).

2.3.3 ORGANIZATION STRUCTURE

PMDA is an independent administrative agency under the Ministry of Health, Labour, and Welfare (MHLW). PMDA is led by a chief executive, who is appointed by the Minister of MHLW (PMDA Act No. 192, 2002). The executive directors of the offices of Review, Safety, and Relief report to the chief executive.

2.3.4 PRIMARY REGULATORY FUNCTIONS

PMDA evaluates the quality, efficacy, and safety of drugs, medical devices, and cellular and tissue-based products in light of current scientific and technological standards. The services provided are shown in Fig. 2.1 (PMDA, 2016a). Companies may seek advice on regulatory submissions through consultations and assistance in the development of clinical trials. The agency assesses compliance with ethical, scientific, and quality standards, including Good Laboratory Practice (GLP), Good Clinical Practice (GCP), Good Postmarketing Study Practice (GPSP), and conducts inspections to ensure that manufacturing facilities are in compliance with good manufacturing practice (GMP), quality management system (QMS), and good gene, cellular, and tissue-based product manufacturing practice (GCTP).

2.3.5 NOTEWORTHY REGULATORY PROVISIONS

There are several important points to keep in mind for companies formulating their regulatory strategies to enter the Japanese market.

2.3.5.1 Special regulatory pathway for regenerative medicine

PMDA has developed a program of Conditional and Time-limited Authorization of Regenerative Medical Products to facilitate and accelerate the development of

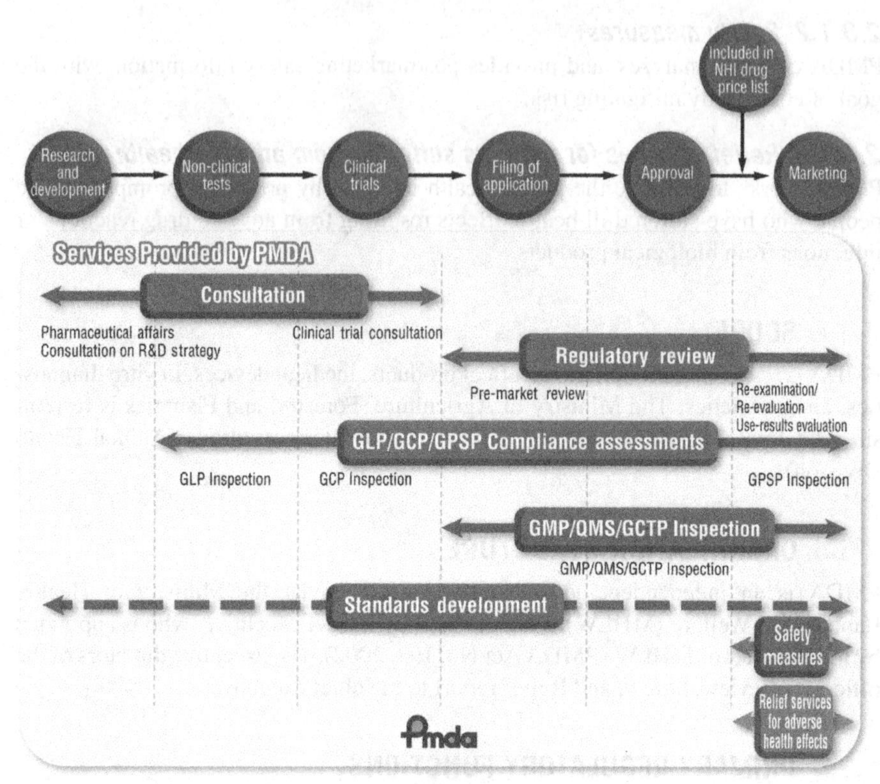

FIGURE 2.1 Outline of PMDA reviews and related services.

regenerative medicine to provide patients suffering from life-threatening diseases with earlier access to potentially effective treatments (PMDA, 2016d; Fig. 2.2).

2.3.5.2 Drug master file system

The *PMDA Act* (*No. 80-61*) provides for a drug master file system (DMF) that allows the manufacturer to submit detailed information about the manufacturing process and intellectual property in support of a regulatory submission without disclosing the information prior to approval (PMDA, 2016e; Fig. 2.3). This provision is especially helpful for foreign manufacturers who rely on local representation by market authorization holder (MAH) to interact with the regulatory agency. In this case, a company can submit DMF directly to PMDA without revealing confidential information to its MAH.

2.3.5.3 Accelerated regulatory pathway for innovative products

PMDA provides a streamlined consultation and review process for products with "*SAKIGAKE*" designation. This pathway is available for innovative medical products (drugs, medical devices, and regenerative therapies) for serious diseases when there is evidence of effectiveness based on nonclinical and early phase clinical studies. This process

Conditional and time-limited authorization
of regenerative medical products
(Gene and cell therapy products)

Conventional regulatory approval process

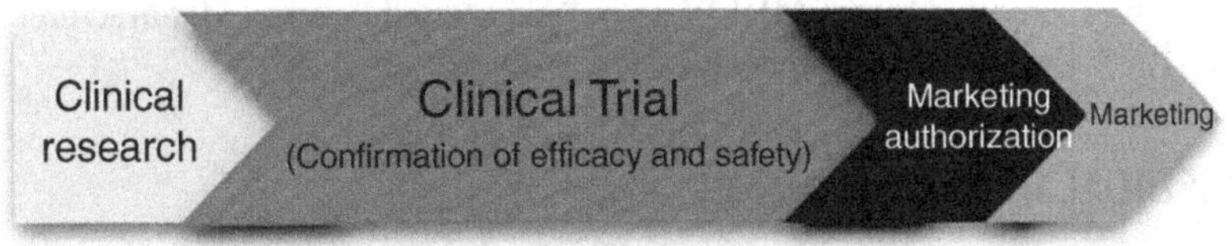

Regulatory system that facilitate early patient access

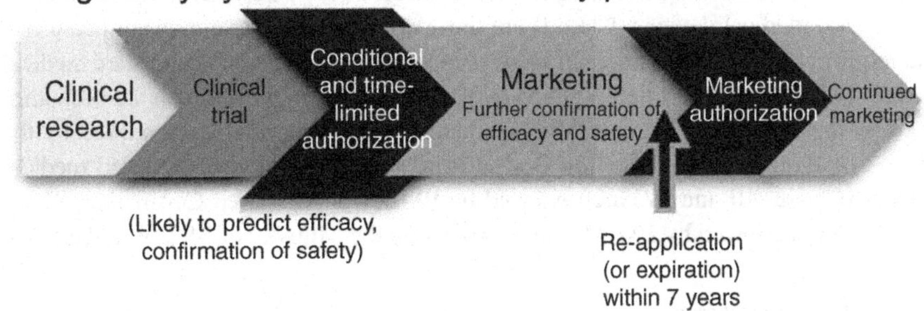

FIGURE 2.2 Conditional and time-limited authorization of regenerative medical products.

Approval reviews for drug products
quoting MF (Simplified outline)

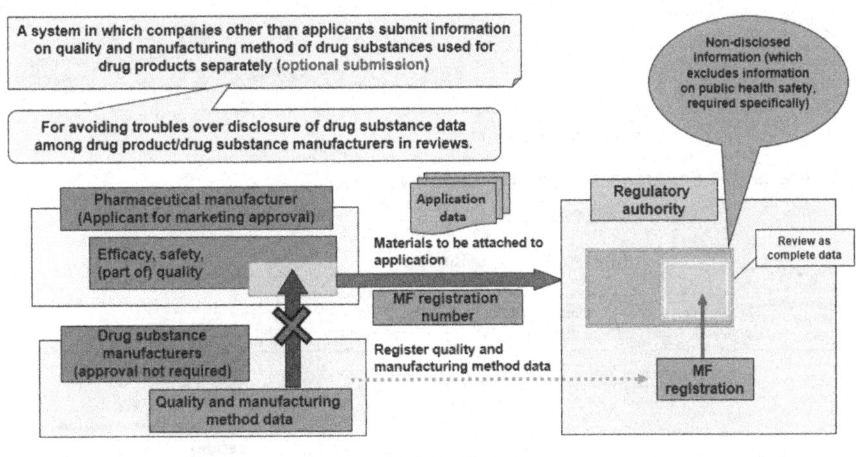

2015/8/19 Pharmaceuticals and medical devices agency (PMDA) 42

FIGURE 2.3 Drug master file.

allows for prioritized regulatory consultation, rolling review during product development, and a shortened review time for market approval (PMDA, 2016f; Fig. 2.4).

2.3.5.4 Accredited foreign manufacturer

A non-Japanese manufacturer (a person/a company) that intends to manufacture drugs, quasidrugs, or medical devices outside of Japan and then export them to Japan is required to be accredited by the MHLW as an "Accredited Foreign Manufacturer" (PMDA, 2016g). This accreditation must be renewed every 5 years.

2.3.6 MEDICAL DEVICES

Similar to regulatory approaches in the United States and Europe, medical devices in Japan are also classified according to risk (PMDA, 2016h; Table 2.6).

General medical devices (Class I) are those considered to pose nonsignificant risk and require conformance with a self-declaration system. Designated controlled medical devices (Class II) are certified by third-party certification bodies, which are preauthorized by the MHLW. Other controlled medical devices (Class II) are reviewed by the PMDA or Registered Certification Body (Third Party). Specially controlled medical devices (Classes III and IV) are reviewed by PMDA or registered Certification Body (Class III) or reviewed by PMDA and approved by the Minister of MHLW (Class IV).

2.3.7 QUASIDRUGS

A quasidrug is defined as a product that has features of both a drug and a cosmetic and has a mild effect on a human. This includes sprays and aerosols. A quasidrug requires an application for approval, consultation, and compliance with GMP.

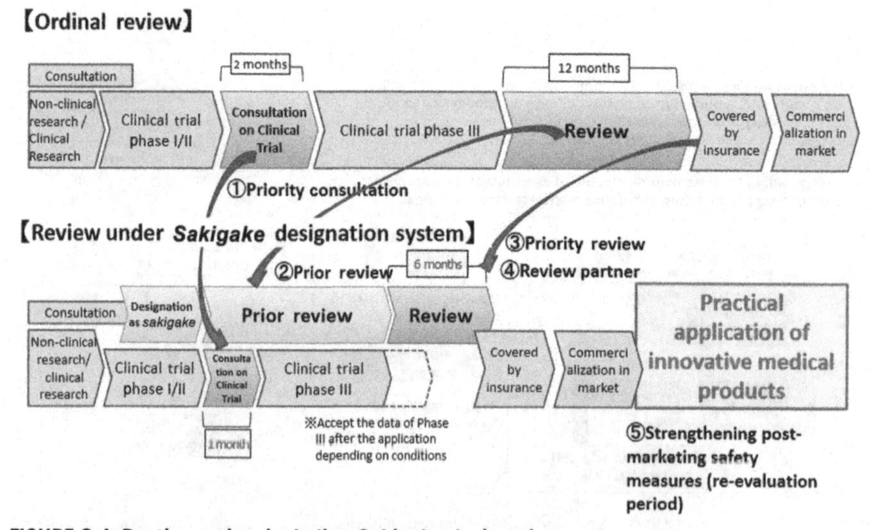

FIGURE 2.4 Routine review including Sakigake designation system.

Table 2.6 Classification and Regulation of Medical Devices in Japan

<table>
<tr><th>International Classification</th><th>Risk Base Medical Device Classification</th><th>Classification</th><th>Risk Level</th><th colspan="2">Type of Regulation</th></tr>
<tr><td>Class 1</td><td>Devices with extremely low risk to the human body in case of problems. Examples: in vitro diagnostic devices, steel-made small devices {including a scalpel, tweezers), X-ray film, devices for dental technique</td><td>General medical device</td><td>Extremely low</td><td colspan="2">Approval/certification not required {Notification/self-declaration)</td></tr>
<tr><td>Class II</td><td>Devices with relatively low risk to the human body in case of problems. Examples: MRI devices, electronic endoscope, catheter for digestive organs, ultrasonic devices, dental alloy</td><td>Controlled medical device</td><td>Low</td><td rowspan="2">Certification by third-party certification (limited to devices for designated controlled medical device, complying with certified standards)</td><td rowspan="2">Approval by the Minister of Health, Labour and Welfare {reviewed by PM DA)</td></tr>
<tr><td>Class III</td><td>Devices with relatively high risk to the human body in case of problems. Examples: Dialyzer, bone prosthesis, mechanical ventilation</td><td rowspan="2">Specially controlled medical device</td><td rowspan="2">Medium/High</td></tr>
<tr><td>Class IV</td><td>Devices highly invasive to patients and with life-threatening risk in case of problems. Examples: pacemaker, artificial cardiac valve, stent graft</td><td colspan="2">Approval by the Minister of Health, Labour and Welfare (reviewed by PM DA)</td></tr>
</table>

2.3.8 COSMETICS

Cosmetics are regulated under the *Pharmaceuticals and Medical Devices Agency Act* (*No. 145-1960*). Manufacturers must submit an application. Cosmetic products are required to comply with certain quality standards, and restrictions on ingredients, labeling, and advertising (MHLW notification 322-1967). PMDA reviews the identity of products and MHLW approves the application.

2.3.9 ORPHAN PRODUCTS

"Priority Review" is the term used to refer to orphan drugs (expected to be used by less than 50,000 patients) and products designated for priority review by the MHLW in consideration of their clinical usefulness and the seriousness of the diseases for which they are indicated.

2.3.10 POSTMARKETING SAFETY MEASURE

Japan has developed a postmarketing structure that places a strong emphasis on safety-related activities, as depicted in Fig. 2.5 which provides an overview of the Japanese Pharmacovigilance Framework (PMDA, 2016i).

Within its PV Framework, the PMDA collects, stores, and analyzes adverse reactions; provides consultations to companies regarding prescription labeling; and provides telephone consultation for consumers regarding questions about drugs and medical devices. PMDA also explores new methods and techniques for postmarketing safety operations, for example, developing new data mining methods and establishing a sentinel medical institution network to detect safety signals. To disseminate information to the public, PMDA makes available on its web site package inserts, drug guides for patients, product recalls, and other urgent safety alert information. Similar to regulatory approaches in the United States and Europe, PMDA requires companies to develop risk management plans for new drugs as well as selected generics.

Regenerative products, which are approved under the new expedited approval scheme, require all patients to be registered. In addition, the company must conduct

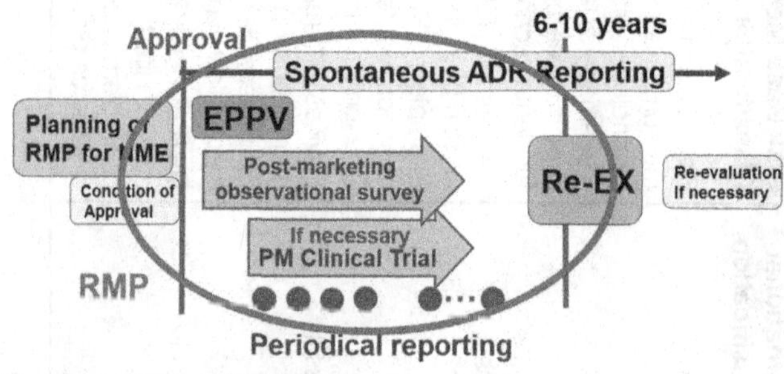

FIGURE 2.5 Pharmacovigilance framework.

a postmarketing study to confirm efficacy and safety before the MAH may reapply for full approval.

2.3.10.1 Early postmarketing phase vigilance

A key feature of the PV Framework is the Early Postmarketing Phase Vigilance (EPPV) (PMDA, 2016j). Under this provision, the MAH is required to closely monitor ADRs in the first 6 months after product launch by intensively collecting ADRs and providing safety information to healthcare providers (HCPs). PMDA collects ADR reports from the MAH and HCPs and holds an expert meeting every 5 weeks to consider possible revision of labeling. In the event that there is a determination of particularly important safety information that requires urgent communication, PMDA requests the MAH to disseminate one of the following two letters which is then posted on the PMDA website and sent to subscribers via email:

1. Yellow letter: Dear Healthcare Professional Letters of Emergent Safety Communications
2. Blue letter: Dear Healthcare Professional Letters of Rapid Safety Communications

2.3.10.2 Mihari project

The MIHARI ("guard") project was started in 2009 to strengthen postmarketing drug safety measures in PMDA by developing a new safety assessment framework using Japanese medical information databases, such as health insurance claims data and medical records (PMDA, 2016k). Pilot studies were created to ensure access to existing electronic healthcare data (EHD) and to develop PV methodologies and techniques to use EHD for quantitative risk evaluation and for evaluating the impact of regulatory actions. This was facilitated by the creation of Medical Information Database Network (MID-NET) that helped to establish access to EHD at 23 hospitals that are part of 10 hub medical institutions.

2.3.10.3 Relief services for adverse health effects

The Relief Fund for Patients Suffering from Adverse Drug Reactions was established in 1979 from ongoing contributions by MAHs to provide compensation to patients or their families for ill health or death as a result of *appropriately* prescribed and purchased drugs (PMDA, 2016a). These include adverse drug reactions, including reactions due to cellular- and tissue-based products, SMON (subacute myelo-optic neuropathy) due to *Clioquinol* (iodochlorhydroxyquin), and ill health as a result of transfusion of blood or blood products contaminated with HIV or Hepatitis C Virus.

2.3.11 REGULATORY SCIENCE

PMDA participates in standard development activities, including Japanese Pharmacopoeia (first published in 1886), Companion Diagnostics Working Group (WG) for biomarkers, and nanomedicine WG.

2.3.12 INTERNATIONAL ACTIVITIES

PMDA collaborates with the Health Authorities of other countries and is an active member of the International Conference on Harmonization of Technical Requirements for Registration of Pharmaceuticals for Human Use (ICH, 2016), International Medical Device Regulators Forum (IMDRF), International Coalition of Medicines Regulatory Authorities (ICMRA), and Harmonization by Doing (HBD).

2.3.13 SUMMARY

Japan has created a comprehensive set of regulations that includes all aspects of drug development, expedited review, orphan drugs, pediatrics, and pharmacovigilance. Japan continues to play a key role in international activities.

2.4 EUROPEAN UNION: EUROPEAN MEDICINES AGENCY

The European Union is an extraordinarily complex political entity as Fig. 2.6 demonstrates (Supranational European and Bodies, 2016). Since the original formation of the European Economic Community, beginning with France, Germany, Italy, Belgium, Netherlands, and Luxemburg in 1958, there have been seven enlargements, to

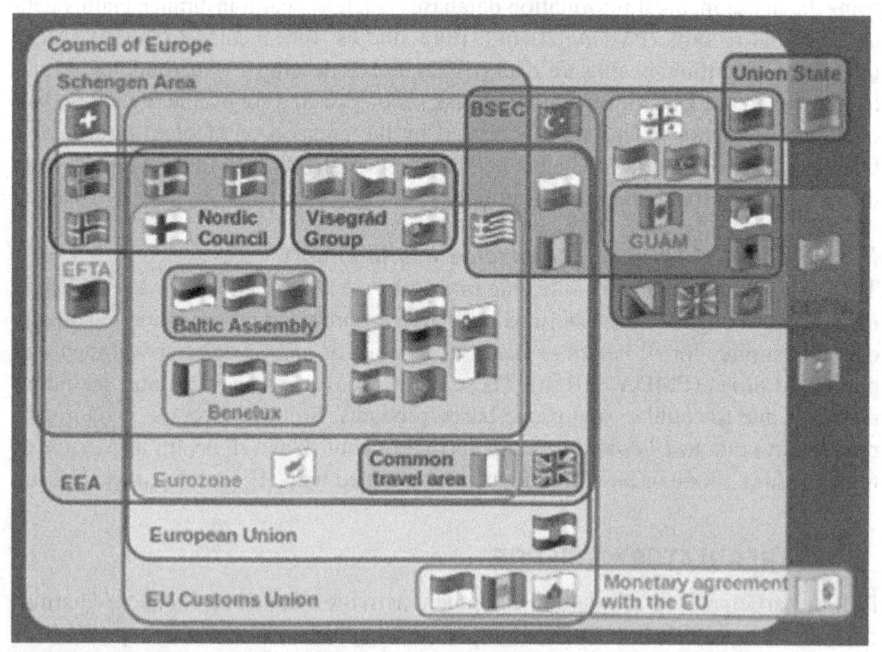

FIGURE 2.6 Relationships between various multinational European organizations and agreements.

include a total of 28 countries, also known as Member States, with Croatia added most recently in 2013, soon to be 27 following Brexit in 2019. Today, there are treaties on commerce, use of common currency, passport-free travel, customs, regulation, and other mutually agreed-upon issues (EU official web site).

This deep historical context and political organizational complexity serve as the backdrop to the development of a regulatory structure for reviewing and approving medicinal products in the EU.

The European Union celebrated 50 years of pharmaceutical regulatory milestones in 2014 (EC, 2016a), marking the anniversary of the Declaration of Helsinki that established Ethical Principles for Clinical Research. The European Medicines Agency (EMA) opened its doors for business in 1995 after decades of gradually increasing centralization processes of its institutions and procedures (EMA, 2016a, 2016b; Table 2.7).

The EMA itself, however, is a structurally decentralized agency of the EU, governed by an independent management board, and managed as a highly networked organization whose activities involve thousands of experts from across Europe who carry out their work through EMA's seven scientific committees (EMA, 2016c, 2016d):

1. Committee for Medicinal Products for Human Use (CHMP)
2. Pharmacovigilance Risk Assessment Committee (PRAC)
3. Committee for Medicinal Products for Veterinary Use (CVMP)
4. Committee for Orphan Medicinal Products (COMP)
5. Committee on Herbal Medicinal Products (HMPC)
6. Committee for Advanced Therapies (CAT)
7. Paediatric Committee (PDCO).

Of particular importance are the CHMP, the CAT, and the PRAC. The CHMP is responsible for preparing the opinion of the Agency on any question relating to the evaluation of medicinal products for human use. The CAT is responsible for preparing the assessment and draft opinions of the Agency regarding applications submitted under the centralized procedure for advanced therapy medicinal products. The PRAC is responsible for assessing all aspects of the risk management of medicines for human use, in particular, in the preparation of recommendations relating to pharmacovigilance activities for medicines for human use and on risk-management systems, including the monitoring of the effectiveness of those risk-management systems. The work of these committees is supported by working parties and other groups.

Table 2.7 Comparison of Key Demographic Features of the EU and the USA

	States	Population	Area	GDP Per Capita
EU	28 (27)[a]	743 (678)[a] million	3.9 (3.8)[a] MM SQ Mi^2	28,800 €
US	50	319 million	3.8 MM SQ Mi^2	53,000 USD

[a]27 states following Brexit, scheduled for completion in 2019.

2.4.1 MISSION AND SCOPE

The mission of the EMA is to foster scientific excellence in the evaluation and supervision of medicines, for the benefit of public and animal health in the European Union (EU) (EMA, 2016e). This is accomplished by

1. facilitating development and access to medicines,
2. evaluating applications for marketing authorization,
3. monitoring the safety of medicines across their lifecycle, and
4. providing information to healthcare professionals and patients.

The EMA does not have regulatory control of the following: the pricing of medicines, patents on medicines, and the availability of medicines, medical devices, homoeopathic medicines, herbal supplements, food supplements, cosmetics, and advertising of medicines.

2.4.2 LEGAL FRAMEWORK

The major legislative activities regarding drugs, biologics, and medical devices are presented in Table 2.8.

Prior to 1995, applications for medicinal products for human use either were reviewed and approved by health authorities of individual countries (nationally approved products) or were approved by health authorities of countries after they had been reviewed and approved earlier by health authorities of other countries (mutual recognition products) (EMA, 2016f).

The creation of the EMA cleared the way for developing a process to centrally approve products, that is, a process which requires but a single application and review that is recognized by all Member States (EMA, 2016g). The EMA provides regulatory control of medicines and medical devices throughout the European Economic Area (EEA), which includes the EU plus three of the four countries of the European Free Trade Association (EFTA): Norway, Iceland, and Liechtenstein, which specifically excludes Switzerland.

EMA regulation encompasses products for human use and veterinary use and the body of EU legislation in the pharmaceutical sector is compiled in Volume 1 (products for human use) and Volume 5 (products for veterinary use). The basic legislation is supported by a compendium of guidelines; those for human use include the following (EC, 2016b):

1. Vol. 2: Notice to applicants and regulatory guidelines for medicinal products
2. Vol. 3: Scientific guidelines for medicinal use
3. Vol. 4: Guidelines for good manufacturing practices for medicinal use
4. Vol. 8: Maximum residue limits
5. Vol. 9: Guidelines for pharmacovigilance for medicinal products
6. Vol. 10: Guidelines for clinical trials.

Standing EMA committees are responsible for managing medicinal products for pediatric use (PDCO), orphan disease populations (COMP), herbal medicinal products (HMPC), and advanced therapies (CAT).

Table 2.8 Key Events and Legislation Impacting Drugs, Biologics, and Medical Devices in EU

Year	Event or Action	Importance
1964	Declaration of Helsinki	Establishes ethical principles for clinical research
1965	Thalidomide disaster	EU decision to authorize medicinal products before permitting access to market
1975	Agreement on joint EU position on market authorization	Establishes first steps for a multistate procedure and a common committee
1983	Agreement of Member States on authorized products	Establishes a uniform way to summarize key characteristics of an authorized product
1987	Concentration Procedure	Requires opinion of EU Level Committee before national authorities authorize innovative products
1990	Council Directive 90/385/EEC	Introduces rules to assure protection of human health with active implantable medical devices
1993	Council Directive 93/42/EEC	Introduces rules to assure protection of human health and safety and proper functioning in single European market
1995	EMA opens for business	Becomes responsible body for authorization of human and veterinary medicinal products through centralized EU-wide procedures
1998	Council Directive 98/79/EC	Introduces rules for in vitro diagnostic medical devices
2001	Directive 2001/20/EC	Establishes Clinical Trial Directive that provides requirements for the conduct of clinical trials
2004	Directive 2001/83/EC, as amended (2004/27/EC)	Introduces rules for copies of biological products
2007	Regulation (EC)1394/2007	Introduces principles and concepts for the regulation of advanced therapy medicinal products based on genes (gene therapy), cells (cell therapy), and tissues (tissue engineering)
2010	Directive 2010/84/EURegulation (EU) No 1235/2010	Strengthens system for safety of medicines: better prevention, detection, and assessment of adverse reactions to medicines, direct patient reporting
2011	Directive 2011/62/EU	Adopts procedures to protect against marketing of falsified medicines, including packaging, EU logo, controls, inspections and record keeping
2012	Joint Plan for Immediate Actions under existing Medical Devices Legislation	Launches plan following the discovery of the fraudulent use of nonmedical grade silicone in breast implants that were manufactured by the company "Poly Implant Prothèse" (PIP)
2012	Proposal for a Regulation of the European Parliament and of the Council on medical devices, and amending Directive 2001/83/EC, Regulation (EC) No 178/2002 and Regulation (EC) No 1223/2009	The revisions affected many kinds of medical devices including in vitro diagnostic medical devices, from home-use items such as sticking plasters, pregnancy tests and contact lenses, to X-ray machines, pacemakers, breast implants, hip replacements and HIV blood tests
2014	Regulation (EU) No 536/2014	Introduces updates to clinical trial directive, in particular, simplification of procedures, improved harmonization, and greater transparency

Article 3 of Regulation (EC) No. 726/2004 defines the scope and eligibility of applications for evaluation under the centralized procedure through which medicinal products must ("mandatory scope") be authorized by the Community. There are also "optional scope" or "generic/hybrid" procedures for certain other products.

2.4.3 CENTRALIZED PROCEDURE

The centralized procedure is compulsory for products using advanced technology and those with selected therapeutic indications.

2.4.3.1 Biotechnology products

Centralized procedure is mandatory for medicinal products developed by means of one of the following biotechnological processes: recombinant DNA technology controlled expression of genes coding for biologically active proteins in prokaryotes and eukaryotes including transformed mammalian cells, hybridoma, and monoclonal antibody methods. Similar biological ("biosimilar") medicinal products which are developed by one of the above biotechnological processes also fall under the mandatory scope of the centralized procedure.

2.4.3.2 Advanced therapies

Advanced therapy medicinal products as defined in Article 2 of Regulation (EC) No. 1394/2007 includes gene therapy medicinal products, somatic cell therapy medicinal products, and tissue engineered products. Advanced therapies have their own set of regulations (EC, 2016c) and must use the centralized procedure.

2.4.3.3 Selected indications

Centralized procedure is mandatory for a new drug which proposes to treat any of the following diseases: acquired immune deficiency syndrome, cancer, neurodegenerative disorder, diabetes, auto-immune diseases, and other auto-immune dysfunctions, and viral diseases.

Products designated as orphan medicinal products are also required to follow the centralized procedure.

2.4.4 CLINICAL TRIALS

On April 16, 2014, the European Commission adopted the new Clinical Trial Regulation (EU No 536/2014), which comes into effect in 2018, with the goals of harmonizing electronic submission and assessment process for clinical trials conducted in multiple Member States and increase transparency of information on clinical trials (EC, 2016d).

2.4.4.1 Investigational medicinal product application

The national competent authorities of the respective Member States are responsible for authorization of clinical trials conducted in their jurisdictions. This includes ensuring that the rights, safety, and well-being of trial subjects are protected and the results of clinical trials are credible.

Routine assessment: The maximum timeframe for the evaluation of an authorization application under the Centralized Procedure is 210 days.

Accelerated assessment: The maximum timeframe can be reduced to 150 days when there is a justification that demonstrates introduction of a new method of therapy or improvement on existing methods, thereby addressing to a significant extent the greater unmet needs for maintaining and improving public health.

Marketing authorization applications that are under the centralized procedure are evaluated by an external system of rapporteur/co-rapporteur/peer reviewer and their assessment team (EMA, 2016h). The agency charges fees, adjusted for inflation, for applications for marketing authorization, and for variations and other changes to marketing authorizations, as well as annual fees for authorized medicines (EMA, 2016i).

2.4.5 MEDICAL DEVICES

The EMA does not have jurisdictional authority over medical devices or in vitro diagnostics. Instead, the European Commission legislated gradual harmonization throughout the EU in the 1990s in parallel to the process for drugs and biologics (EC, 2016e, 2016f, 2016g; Table 2.2: 1990, 1993, 1998) and subsequently established a process involving Notified Bodies (EC, 2016h). A Notified Body is an organization designated by an EU country to assess the conformity of certain products before being placed on the market. These bodies carry out tasks related to conformity assessment procedures set out in the applicable legislation, when a third party is required. The European Commission publishes a list of such notified bodies.

Governance is also provided through the EU Competent Authorities for Medical Devices (CAMD) which, following the breast implant scandal in 2012 that involved the fraudulent use of nonmedical grade silicone in the manufacture of breast implants by "Poly Implant Prothèse" (PIP), carried out a PIP Action Plan to enhance further collaboration, market surveillance, and communication regarding medical devices across Europe (EC, 2016i).

2.4.6 PHARMACOVIGILANCE

The pharmacovigilance system in the EU is a collaborative activity among the regulatory authorities in Member States, the European Commission and the EMA (EMA, 2016j). In some Member States, regional centers are in place under the coordination of the national Competent Authority. The EMA's role is to coordinate the EU pharmacovigilance system and to operate specific systems, services, and processes as laid down in the EU legislation (EMA, 2016k). In building out the PV systems, EMA developed a series of good pharmacovigilance practice (GVP) modules (EMA, 2016l):

1. Pharmacovigilance systems and quality systems
2. Pharmacovigilance system master file
3. Pharmacovigilance inspections
4. Pharmacovigilance system audits
5. Risk management systems
6. Management and reporting of adverse reactions
7. Periodic safety update reports
8. Postauthorisation safety studies
9. Signal management
10. Additional monitoring
11. Void (planned topics have been addressed by other documents.)
12. Void (planned topics have been addressed by other documents.)
13. Void (planned topics have been addressed by other documents.)
14. Void (planned topics have been addressed by other documents.)
15. Safety communication
16. Risk-minimization measures: selection of tools and effectiveness indicators.

2.4.6.1 Risk management plan (EMA, 2016m)

A medicine is authorized after it has been shown that the data provided by the Marketing Authorization Applicant demonstrate that its benefits outweigh its risks for the target population. However, not all potential or actual adverse reactions are identified by the time an initial marketing authorization is granted. The aim of risk management is to address uncertainties in the safety profile at different points in the lifecycle, and to plan accordingly.

Marketing authorization applicants are required to submit risk management plans, which include information on a medicine's safety profile and plans for pharmacovigilance activities designed to gain greater knowledge. They also explain how risks will be minimized in patients and how those efforts will be measured.

2.4.6.2 Qualified person of pharmacovigilance (QPPV)

In 2001, the EU created the position of Qualified Person of Pharmacovigilance (QPPV) (EMA, 2016n). The QPPV serves as a single point of contact concerning safety of marketed products and PV inspections and ensures that a system is in place to be continuously available to be contacted by regulatory authorities in the European economic area (EEA) [the 28 Member States, plus three countries of the European Free Trade Association: Norway, Iceland, and Liechtenstein; this explicitly excludes Switzerland].

Among the many other important functions are the following: establishment and maintenance of a PV system, access to the PV system master file, overview of medicinal product safety profiles, full and timely response to requests from competent authorities, awareness of inspection and audit findings, assurance of a valid and operational database, and assurance of an adequate disaster recovery plan for the continuance of PV activities. Many other jurisdictions have emulated this approach to management of pharmacovigilance activities.

2.4.7 SUMMARY

The EU has created a robust, highly centralized, regulatory system for medicinal and medical device products through a complex set of Directives and Regulations that govern 28 Member States of the European Economic Area (to be reduced to 27 following Brexit in 2019). Most of these activities are managed through the EMA, though medical devices are largely evaluated through independent Notified Bodies. The EMA has also created a highly integrated set of documents and established the QPPV as the single point of contact concerning the safety of marketed products.

2.5 ROLE OF CIOMS AND ICH

There are numerous organizations working to foster collaboration and harmonization across international biomedical communities. Of them, CIOMS and ICH are chosen as model organizations to be examined below.

2.5.1 CIOMS

The Council for International Organizations of Medical Sciences (CIOMS) is an international, nongovernmental, nonprofit organization established jointly by WHO and UNESCO in 1949 (CIOMS, 2016). CIOMS has representation from a substantial proportion of the biomedical scientific community, including international, national, and associate member organizations of many biomedical disciplines, national academies of sciences, and medical research councils.

CIOMS has had a significant impact on pharmacovigilance and the evolution of risk management frameworks. From 1990 to 2016, they have developed the following work products:

1. International reporting of adverse drug reactions
2. International reporting of periodic drug–safety summaries
3. Guidelines for preparing core clinical-safety information on drugs
4. Benefit–risk balance for marketed drugs: evaluating safety signals
5. Current challenges in pharmacovigilance: pragmatic approaches
6. Management of safety information from clinical trials
7. The development safety update report (DSUR): harmonizing the format and content for periodic safety reporting during clinical trials
8. Practical aspects of signal detection in pharmacovigilance
9. Practical approaches to risk minimization for medicinal products
10. Considerations for applying good meta-analysis practices to clinical safety data within the biopharmaceutical regulatory process.

2.5.2 ICH

The International Conference (now Council) on Harmonization (ICH, 2016) began in 1990 as a natural extension of the years of experience with harmonization activities by the EU (ICH, 2016). Harmonization activities cover all aspects of drug development and fall into the following QSEM scheme:

1. Q: Quality guidelines, including GMP.
2. S: Safety guidelines, including carcinogenicity, genotoxicity, and reprotoxicity.
3. E: Efficacy guidelines, including the design, conduct, safety, and reporting of clinical trials; novel types of medicines derived from biotechnological processes and the use of pharmacogenetics/genomics techniques to produce better targeted medicines.
4. M: Multidisciplinary guidelines, including cross-cutting topics which do not easily fit into one of the other categories: medical terminology (MedDRA), the common technical document (CTD), and the development of electronic standards for the transfer of regulatory information (ESTRI).

2.6 OVERALL SUMMARY

Regulatory oversight of medical products began with the FDA and evolved gradually during its early history in response to events that occurred in the United States. When the thalidomide tragedy occurred in Europe, it triggered a fundamental shift toward increased regulatory expectations and requirements for the entire drug development enterprise. Although the transformation began in the United States, the thalidomide event also contributed to Europe's efforts to harmonize, then centralize, its individual country regulations. Likewise, Japan developed its own regulatory apparatus in response to the significant adverse events that occurred within its borders and beyond. All three regions benefited from the development of infrastructural concepts created by CIOMS and facilitated by ICH. These international harmonization efforts are expected to continue but the discussions will need to be more inclusive to accommodate the evolving global landscape surrounding the development, manufacture, and commercialization of biomedical products.

REFERENCES

Bachorik, L. (2016). *FDA's regulatory framework—An overview*. Retrieved from http://www.fda.gov/downloads/internationalprograms/newsevents/ucm414990.pdf.

CIOMS. (2016). Retrieved from http://www.cioms.ch/index.php/2012-06-07-19-16-08/about-us.

DHHS. (2016). Department Health & Human Services. *Organizational chart*. Retrieved from https://www.hhs.gov/about/agencies/orgchart/.

EC. (2016a). *50 Years: EU pharmaceutical regulation milestones*. Retrieved from https://ec.europa.eu/health/sites/health/files/human-use/50years/docs/50years_pharma_timeline_v3.pdf.

EC. (2016b). *Eudralex—EU Legislation*. Retrieved from http://ec.europa.eu/health/documents/eudralex_en.

EC. (2016c). *Advanced therapies regulation*. Retrieved from http://ec.europa.eu/health/human-use/advanced-therapies/index_en.htm.

EC. (2016d). *New Clinical Trial - Regulation* EU No 536/2014. Retrieved from http://ec.europa.eu/health/human-use/clinical-trials/regulation/index_en.htm.

EC. (2016e). *Medical devices.* Retrieved from http://ec.europa.eu/growth/sectors/medical-devices_en.

EC. (2016f). *Medical device regulatory framework.* Retrieved from http://ec.europa.eu/growth/sectors/medical-devices/regulatory-framework_en.

EC. (2016g). *Revisions of medical device directives.* Retrieved from http://ec.europa.eu/growth/sectors/medical-devices/regulatory-framework/revision_en.

EC. (2016h). *Notified bodies.* Retrieved from http://ec.europa.eu/growth/single-market/goods/building-blocks/notified-bodies_en.

EC. (2016i). *PIP action plan.* Retrieved from http://ec.europa.eu/growth/sectors/medical-devices/pip-action-plan_en.

EMA. (2016a). *Home.* Retrieved from http://www.ema.europa.eu/ema/index.jsp?curl=pages/home/Home_Page.jsp&mid=.

EMA. (2016b). *History of EMA.* Retrieved from http://www.ema.europa.eu/ema/index.jsp?curl(pages/about_us/general/general_content_000628.jsp&mid(WC0b01ac058087addd.

EMA. (2016c). *EMA How We Work.* Retrieved from http://www.ema.europa.eu/ema/index.jsp?curl(pages/about_us/general/general_content_000125.jsp&mid(WC0b01ac0580028a46.

EMA. (2016d). *EMA committees.* Retrieved from http://www.ema.europa.eu/ema/index.jsp?curl(pages/about_us/general/general_content_000217.jsp&mid=.

EMA. (2016e). *What we do.* Retrieved from http://www.ema.europa.eu/ema/index.jsp?curl(pages/about_us/general/general_content_000091.jsp.

EMA. (2016f). *Authorization of medicines.* Retrieved from http://www.ema.europa.eu/ema/index.jsp?curl(pages/about_us/general/general_content_000109.jsp&mid(WC0b01ac0580028a47.

EMA. (2016g). *European Medicines Agency pre-authorisation procedural advice for users of the centralised procedure.* Retrieved from http://www.ema.europa.eu/docs/en_GB/document_library/Regulatory_and_procedural_guideline/2009/10/WC500004069.pdf.

EMA. (2016h). *CHMP rapporteur/co-rapporteur/peerreviewer appointment in the centralized procedure. SOP.* Retrieved from http://www.ema.europa.eu/docs/en_GB/document_library/Standard_Operating_Procedure_-_SOP/2009/09/WC500002991.pdf.

EMA. (2016i). *Fee structure.* Retrieved from http://www.ema.europa.eu/ema/index.jsp?curl(pages/regulation/document_listing/document_listing_000327.jsp.

EMA. (2016j). *Pharmacovigilance.* Retrieved from http://www.ema.europa.eu/ema/index.jsp?curl(pages/regulation/general/general_content_000258.jsp&mid(WC0b01ac0580b18c76.

EMA. (2016k). *Pharmacovigilance legislation.* Retrieved from http://www.ema.europa.eu/ema/index.jsp?curl(pages/special_topics/general/general_content_000491.jsp.

EMA. (2016l). *Good pharmacovigilance practices (GVP).* Retrieved from http://www.ema.europa.eu/ema/index.jsp?curl(pages/regulation/document_listing/document_listing_000345.jsp&mid(WC0b01ac058058f32c.

EMA. (2016m). *Risk management plan.* Retrieved from http://www.ema.europa.eu/ema/index.jsp?curl(pages/regulation/general/general_content_000683.jsp&mid(WC0b01ac058067a113.

EMA. (2016n). *Qualified person PV (QPPV).* Retrieved from http://www.ema.europa.eu/docs/en_GB/document_library/Scientific_guideline/2012/06/WC500129132.pdf.

FDA. (2016a). *FDA fundamentals*. Retrieved from http://www.fda.gov/AboutFDA/Transparency/Basics/ucm192695.htm.

FDA. (2016b). *Legislation*. Retrieved from http://www.fda.gov/RegulatoryInformation/Legislation/.

FDA. (2016c). *What does FDA regulate?* Retrieved from http://www.fda.gov/AboutFDA/Transparency/Basics/ucm194879.htm.

FDA. (2016d). *History*. Retrieved from http://www.fda.gov/AboutFDA/WhatWeDo/History/default.htm.

FDA. (2016e). *Drug Price Competition and Patent Term Restoration Act of 1984. Statement of Daniel Troy, Chief Counsel*, FDA, August 1, 2003. Retrieved from http://www.fda.gov/newsevents/testimony/ucm115033.htm.

FDA. (2016f). *FDA Backgrounder on FDAMA*. Retrieved from http://www.fda.gov/RegulatoryInformation/Legislation/SignificantAmendmentstotheFDCAct/FDAMA/ucm089179.htm.

FDA. (2016g). *Best Pharmaceuticals for Children Act and Pediatric Research Equity Act*. Retrieved from http://www.fda.gov/ScienceResearch/SpecialTopics/PediatricTherapeuticsResearch/ucm509707.htm.

FDA. (2016h). *Fact sheet: Pediatric provisions in the Food and Drug Administration Safety and Innovation Act (FDASIA)*. Retrieved from http://www.fda.gov/RegulatoryInformation/Legislation/SignificantAmendmentstotheFDCAct/FDASIA/ucm311038.htm.

FDA. (2016i). *Mission*. Retrieved from http://www.fda.gov/aboutfda/whatwedo/default.htm.

FDA. (2016j). *What does FDA do*? Retrieved from http://www.fda.gov/AboutFDA/Transparency/Basics/ucm194877.htm.

FDA. (2016k). *FDA organization*. Retrieved from http://www.fda.gov/AboutFDA/CentersOffices/default.htm.

FDA. (2016l). *Commissioner's office: Organization chart*. Retrieved from http://www.fda.gov/aboutfda/centersoffices/organizationcharts/ucm393155.htm.

FDA. (2016m). 21 USC FD&CA *Chapter V: Drugs and devices*. Retrieved from http://www.fda.gov/RegulatoryInformation/Legislation/FederalFoodDrugandCosmeticActFDCAct/FDCActChapterVDrugsandDevices/default.htm.

FDA. (2016n). The 1938 Food, Drug, and Cosmetic Act. Retrieved from http://www.fda.gov/AboutFDA/WhatWeDo/History/ProductRegulation/ucm132818.htm.

FDA. (2016o). *Frequently asked questions about therapeutic biological products*. Retrieved from http://www.fda.gov/drugs/developmentapprovalprocess/howdrugsaredevelopedandapproved/approvalapplications/therapeuticbiologicapplications/ucm113522.htm.

FDA. (2016p). 42 USC §262 *Regulation of biological products*. Retrieved from http://www.fda.gov/RegulatoryInformation/Legislation/ucm149278.htm.

FDA. (2016q). 42 USC 351(k) (*biosimilar or interchangeable*). Retrieved from http://www.fda.gov/Drugs/GuidanceComplianceRegulatoryInformation/Guidances/ucm259806.htm.

FDA. (2016r). 21 CFR. Retrieved from http://www.fda.gov/MedicalDevices/DeviceRegulationandGuidance/Databases/ucm135680.htm.

FDA. (2016s). 21 CFR § 312 *IND regulations*. Retrieved from http://www.accessdata.fda.gov/scripts/cdrh/cfdocs/cfcfr/CFRsearch.cfm?CFRPart=312.

FDA. (2016t). *Regulations regarding good clinical trial practices*. Retrieved from http://www.fda.gov/ScienceResearch/SpecialTopics/RunningClinicalTrials/ucm155713.htm.

FDA. (2016u). 21 CFR § 812. Retrieved from https://www.accessdata.fda.gov/scripts/cdrh/cfdocs/cfCFR/CFRSearch.cfm?CFRPart=812.

FDA. (2016v). 21 CFR § 314 *NDA regulations*. Retrieved from http://www.accessdata.fda.gov/scripts/cdrh/cfdocs/cfcfr/CFRsearch.cfm?CFRPart=314.

FDA. (2016w). *Guidance for industry: Expedited programs for serious conditions—drugs and biologics*. Retrieved from http://www.fda.gov/downloads/drugs/guidancecomplianceregulatoryinformation/guidances/ucm358301.pdf.

FDA. (2016x). *Information for industry (biosimilars)*. Retrieved from http://www.fda.gov/MedicalDevices/DeviceRegulationandGuidance/Databases/ucm135680.htm.

FDA. (2016y). 21 CFR § 600. *Biological products: General*. Retrieved from http://www.accessdata.fda.gov/scripts/cdrh/cfdocs/cfcfr/CFRsearch.cfm?CFRPart=600.

FDA. (2016z). 21 CFR § 601. *Licensing*. Retrieved from http://www.accessdata.fda.gov/scripts/cdrh/cfdocs/cfcfr/CFRSearch.cfm?CFRPart=601.

FDA. (2016aa). 21 CFR § 610. *General biological products standards*. Retrieved from http://www.accessdata.fda.gov/scripts/cdrh/cfdocs/cfcfr/CFRSearch.cfm?CFRPart=610.

FDA. (2016bb). *Regulatory controls for medical devices*. Retrieved from http://www.fda.gov/medicaldevices/deviceregulationandguidance/overview/generalandspecialcontrols/ucm2005378.htm.

FDA. (2016cc). *Kefauver–Harris Amendments*. Retrieved from http://www.fda.gov/forconsumers/consumerupdates/ucm322856.htm.

FDA. (2016dd). *REMS FDA basics webinar: A brief overview of risk evaluation and mitigation strategies (REMS)*. Retrieved from http://www.fda.gov/aboutfda/transparency/basics/ucm325201.htm.

FDA. (2016ff). *Chronological History of the Center for Veterinary Medicine*. Retrieved from http://www.fda.gov/aboutfda/whatwedo/history/forgshistory/cvm/ucm142587.htm.

FDA. (2016gg). *CFSAN*. Retrieved from http://www.fda.gov/AboutFDA/CentersOffices/OrganizationCharts/ucm411071.htm.

FDA. (2016ii). *Tobacco Control Act*. Retrieved from http://www.fda.gov/TobaccoProducts/Labeling/RulesRegulationsGuidance/ucm246129.htm.

ICH. (2016). Retrieved from http://www.ich.org/home.html.

Janssen, W. F. (1981). *The story of the laws behind the labels*. Retrieved from http://www.fda.gov/AboutFDA/WhatWeDo/History/Overviews/ucm056044.htm.

PMDA. (2016a). *Outline of relief services*. Retrieved from http://www.pmda.go.jp/english/relief-services/0002.html.

PMDA. (2016b). *Act*. Retrieved from http://law.e-gov.go.jp/htmldata/H14/H14HO192.html.

PMDA. (2016c). *Outline of reviews*. Retrieved from https://www.pmda.go.jp/english/review-services/outline/0001.html.

PMDA. (2016d). *Conditional and time-limited authorization of regenerative medical products*. Retrieved from http://www.pmda.go.jp/files/000211767.pdf.

PMDA. (2016e). *Drug master file*. https://www.pmda.go.jp/english/review-services/reviews/mf/0001.html.

PMDA. (2016f). *Sakigake*. Retrieved from http://www.mhlw.go.jp/english/policy/health-medical/pharmaceuticals/dl/140729-01-02.pdf.

PMDA. (2016g).*Pharmaceutical affairs law*. Retrieved from http://www.japaneselawtranslation.go.jp/law/detail/?ft(1&re(02&dn(1&x(30&y(11&co(01&ia(03&ky(pharmaceutical(affairs(law&page(1.

PMDA. (2016h). *Med devices*. Retrieved from http://www.std.pmda.go.jp/stdDB/index_e.html.

PMDA. (2016i). *Outline of PV framework*. Retrieved from http://www.pmda.go.jp/files/000211766.pdf.

PMDA. (2016j). *Outline of PM safety measures*. Retrieved from https://www.pmda.go.jp/english/safety/outline/0001.html.

PMDA. (2016k). *MIHARI*. Retrieved from http://www.pmda.go.jp/english/safety/surveillance-analysis/0001.html.

Supranational European Bodies. (2016). *Supranational European bodies*. Retrieved from https://upload.wikimedia.org/wikipedia/commons/thumb/1/1a/Supranational_European_Bodies-en.svg/2000px-Supranational_European_Bodies-en.svg.png.

FURTHER READINGS

European Union Official Website. (2016). *European Union Official Website*. Retrieved from https://europa.eu/european-union/index_en.

FDA. (2016ee). *Office of Surveillance and Epidemiology*. Retrieved from http://www.fda.gov/AboutFDA/CentersOffices/OfficeofMedicalProductsandTobacco/CDER/ucm106491.htm.

FDA. (2016hh). Cosmetics. Retrieved from http://www.fda.gov/Cosmetics/default.htm.

CHAPTER

Drugs

3

Daniela Drago*, Eunjoo Pacifici, Susan Bain****

**George Washington University, Washington, DC, United States; **International Center for Regulatory Science, University of Southern California, Los Angeles, CA, United States*

"When I woke up just after dawn on September 28, 1928, I certainly didn't plan to revolutionize all medicine by discovering the world's first antibiotic, or bacteria killer, ... But I guess that was exactly what I did."

– Alexander Fleming

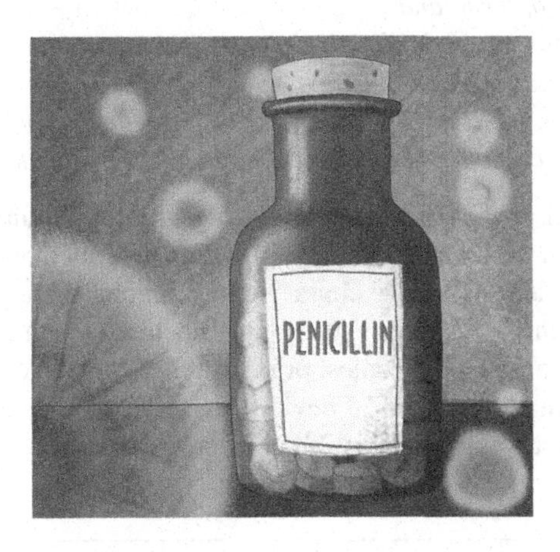

An Overview of FDA Regulated Products. http://dx.doi.org/10.1016/B978-0-12-811155-0.00003-X

CHAPTER OBJECTIVES

After reading this chapter, the reader will be able to:

- identify what constitutes a drug,
- identify the three major categories of drugs,
- describe the regulatory pathways to commercialization for each drug category,
- identify the objectives and elements of nonclinical research and the phases of clinical research, and
- describe what is meant by bioequivalence.

3.1 IS THE PRODUCT A DRUG?

A drug is a substance that produces an effect on the structure or function of the body by a chemical action or by being metabolized. The federal law that sets out the requirements for these types of products in the United States is the "Federal Food, Drug, and Cosmetic Act of 1938 (FD&C Act)" and its associated amendments. The FD&C Act defines "drugs" as:

> *articles recognized in the official United States Pharmacopoeia, official Homoeopathic Pharmacopoeia of the United States, or official National Formulary, or any supplement to any of them; and*
> *articles intended for use in the diagnosis, cure, mitigation, treatment, or prevention of disease in man or other animals; and*
> *articles (other than food) intended to affect the structure or any function of the body of man or other animals (Federal Food Drug and Cosmetic Act of 1938).*

Under the law, all medical products (i.e., drugs, biologics, and medical devices) are intended for use in the diagnosis, cure, mitigation, treatment, or prevention of disease. However, there are distinctive characteristics among these product categories. Drugs are small organic molecules usually manufactured using chemical synthesis. Biologics are large, complex molecules, or mixtures of molecules, that are generally derived from living material. Medical devices are products that do not achieve their intended purpose through chemical action in or on the body.

3.2 HOW ARE DRUGS CATEGORIZED?

In the United States, drug products are categorized as new drugs, generics, or over-the-counter drugs based on their novelty and level of risk.

A *new drug* is one that has never been approved by the Food and Drug Administration (FDA). It may have many potential unknown risks and benefits. Therefore, an applicant[1] who wants to bring a new drug to the market needs to submit a regulatory application and receive approval from the FDA. This application, called the New Drug Application (NDA), contains an extensive set of data that demonstrates the quality, safety, and efficacy of the drug. Specifically, the application includes the analysis of the physicochemical properties of the drug as well as results from studies in animals and in humans. The process of bringing a new drug to the market is long, costly, and has a high failure rate. However, patent protection and market exclusivities provide incentives to applicants and sponsors[2] that undertake drug development.

A *generic drug* is a "copy" of a brand drug. Typically, it has the same indication, active ingredient(s), dosage form, strength, and route of administration. An active ingredient, also called a drug substance, is the component of the drug that provides pharmacological activity. By the time the patent of a brand drug has expired, the drug has usually been on the market for many years and its safety and efficacy profile is well known. FDA only requires that the generic version of the drug demonstrates that it is bioequivalent to the original brand version. In the United States, an applicant wanting to market a generic drug submits an Abbreviated New Drug Application (ANDA). This pathway is used when a drug has the same indication, active ingredient(s), dosage form, strength, and route of administration as the brand drug.

In addition to the NDA and ANDA pathways, applicants also can use a hybrid pathway called the 505(b)2 Application. This pathway is used if a drug contains the same active ingredient as a brand product but is not identical in dosage form, strength, route of administration, and/or indication.

An *over-the-counter (OTC) drug* can be purchased for self-administration without a prescription or the supervision of a healthcare professional. An applicant can develop and market an OTC drug in the United States if the product meets the requirements found in an OTC drug monograph. An OTC drug monograph is developed for each category of OTC drugs and is published in the Federal Register. OTC drug monographs specify ingredients, indications, and other labeling requirements (e.g., doses, acceptable formulations, and instructions for use). If an applicant wants to market an OTC drug that does not have a monograph, it must use a new drug approval process pathway.

A *homeopathic drug* is recognized by the FD&C Act but is not regulated like conventional drugs and biologics. The practice of homeopathy is based on the concept that disease symptoms can be treated by small doses, often highly diluted, of ingredients that cause similar symptoms in healthy subjects. For example, the diluted extract of onion

---

[1]An applicant can be an individual, a company, an agency, an academic institution, a private organization, or any other organization. The term "applicant" is defined in 21 CFR 314.3(b): "Applicant means any person who submits an application or abbreviated application or an amendment or supplement to them under this part to obtain FDA approval of a new drug or an antibiotic drug and any person who owns an approved application or abbreviated application."

[2]The term "sponsor" is defined in 21 CFR 312.3(b): "Sponsor means a person who takes responsibility for and initiates a clinical investigation." Generally, the sponsor and the applicant are the same organization. However, this is not always the case.

may be used for a common cold and hay fever because it is known to cause tearing of the eyes and dripping of the nose in healthy individuals. Homeopathic products are not required to demonstrate safety or efficacy before being allowed on the market. The FDA has exempted homeopathic drug products from putting expiration dates and the identity and strength of each active ingredient on their labels. These products instead follow the standards for strength, quality, and purity set forth in the Homeopathic Pharmacopeia.

Fig. 3.1 shows a decision tree for the categorization of a drug.

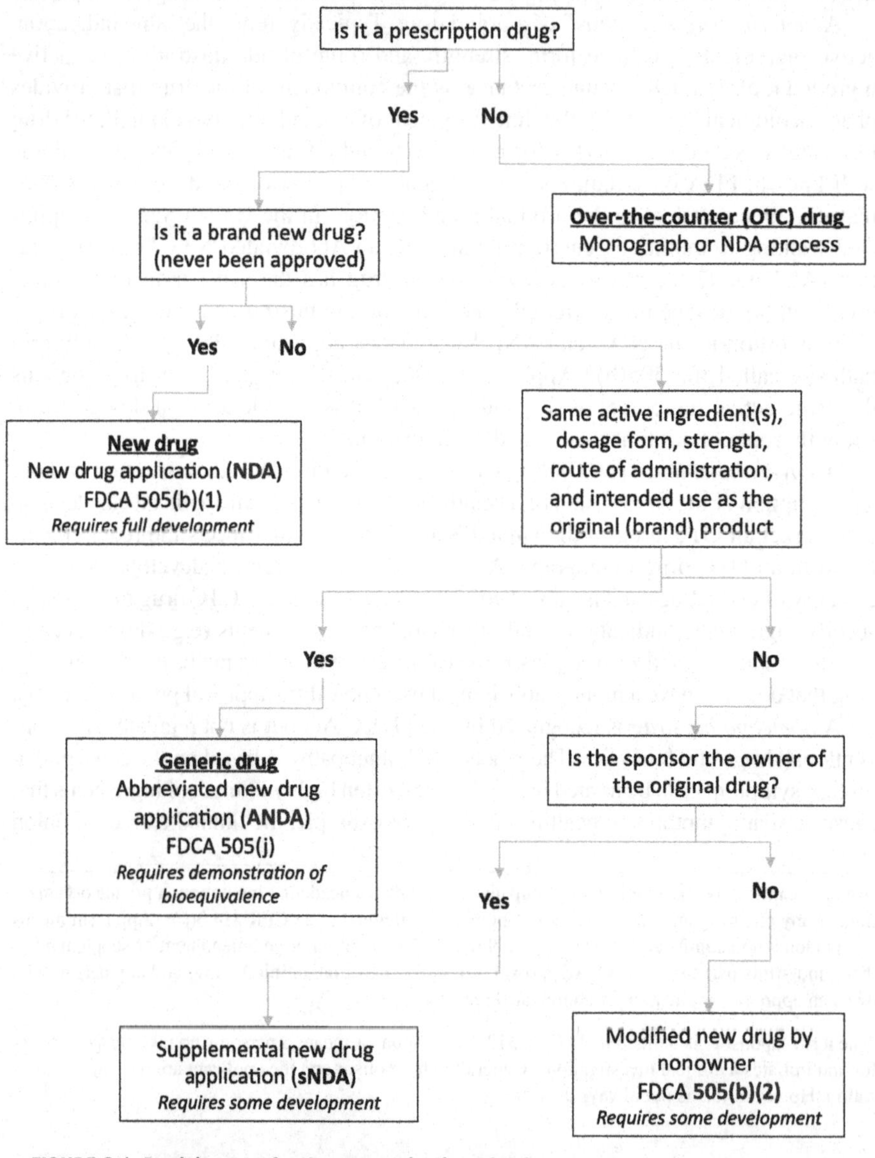

FIGURE 3.1 Decision tree for the categorization of a drug.

3.3 WHAT IS A REGULATORY STRATEGY AND WHAT DOES IT ENTAIL?

A regulatory strategy is a roadmap that allows an applicant to move its regulated product from development to commercialization. It takes into account the applicant's objectives (both short- and long-term), the current and projected regulatory environment, and the needs of multiple stakeholders within and outside the organization. A regulatory strategy typically incorporates background information, the development plan (including the target product profile), regulations and guidance documents, outstanding issues or questions, past precedents (if any), and recommendations.

3.3.1 THE TARGET PRODUCT INDICATION

For drug developers, an important factor to consider is whether there are other similar products with the same target indication already on the market. If so, the applicant should review the pertinent information about these products such as their approved labeling and their Summary Basis of Approval (SBA). The SBA contains a summary of the data and information that FDA evaluated during the drug approval process. The review of publicly available data provides insights into the outcome measures acceptable to regulators, patients, healthcare providers, and payers. Individuals and organizations can also obtain information on products currently in clinical development through public databases such as the US National Institutes of Health's clinicaltrials.gov.

3.3.2 THE TARGET PATIENT POPULATION

Identifying the target patient population allows the applicant to investigate how the patients are distributed geographically. This will help to evaluate where clinical trials should be conducted. An applicant should also consider whether there are specific populations that have a high incidence of the target disease. In addition, it is important to determine whether the target disease is a rare condition. In that case, the drug might qualify for orphan drug designation. In the United States, orphan drug designation is granted to a drug that targets a disease afflicting fewer than 200,000 people, or one that will not be profitable within 7 years following FDA approval. Benefits of an orphan drug designation include a 7-year market exclusivity, development and regulatory assistance from the FDA, access to the clinical research grant program, a 50% tax credit for the cost of clinical research, and a waiver of FDA user fees.

3.3.3 THE FDA-EXPEDITED PROGRAMS

Identifying whether the product qualifies for an expedited program allows the applicant to explore one of the FDA-accelerated pathways to the US market. The FDA developed four approaches designed to expedite access to promising therapeutics. These programs are available for drugs that are intended to treat serious diseases and fill an unmet medical need. The programs are described in Section 3.5.2 of this chapter.

3.3.4 THE TARGET MARKETS

The applicant needs to consider as early as possible where the drug will be marketed. If the objective is to commercialize the drug only in the United States, then complying with FDA regulations is sufficient. However, if the organization wants to commercialize the drug internationally, the applicant needs to develop a global product development strategy that addresses the relevant regional or national regulatory requirements. For example, some countries like China and India require clinical trials to be conducted on their local populations.

3.4 WHAT ARE THE KEY STEPS OF A NEW DRUG DEVELOPMENT PROCESS?

The path from understanding a disease to bringing a new medicine to the market is long, difficult, and expensive. The journey starts when a connection is made between a disease process and a biological target. From that point, drug developers conduct extensive laboratory and animal tests of natural and synthetic compounds to identify the most promising candidate for clinical development. As a molecule moves from the laboratory into the clinic for human testing, the process becomes increasingly regulated.

It typically takes 12–15 years and 1.3–2.6 billion dollars to develop a new drug (DiMasi, Grabowski, & Hansen, 2016). The process is usually divided into six main stages: the discovery and preclinical stages followed by phases 1, 2, 3, and 4 clinical trials. In the United States, any individual or organization that wants to start conducting a clinical trial with a new drug must demonstrate that the product will be safe enough to be tested on humans. This is done through the submission of an Investigational New Drug (IND) application to the FDA and of an Institutional Review Board (IRB) submission to an ethical review board. An IND application contains data from pharmacology and toxicology studies, information on manufacturing and controls, previous human experience (if any), and information on the proposed clinical trial. If the FDA does not object to the information contained in the IND application within 30 days of receipt, the sponsor can initiate trials in human subjects. As development continues, the sponsor will proceed to collect data and determine whether the drug is safe and effective. The FDA remains involved throughout the process with its oversight and guidance.

To illustrate the development and approval of a new drug, the following examples are used: lixisenatide, a drug for treating type 2 diabetes mellitus, and rucaparib, a drug for treating ovarian cancer. The two drugs, both approved in 2016 in the United States, employed different regulatory strategies according to their product characteristics and indications of use.

3.4.1 THE DISCOVERY OF A DRUG CANDIDATE

The first step in the drug development process is to discover biological targets that can be used for the diagnosis, cure, mitigation, treatment, or prevention of a disease. A target can be a specific receptor, enzyme, or gene that, when modulated, influences a biological function associated with a disease.[3] The drug screening process consists of testing many compounds in assays relevant to the disease of interest. If a compound or its structural derivatives continue to show promise, they are considered as leads and enter the drug development process. At the end of the discovery step, the lead compound is selected and needs to be formulated into a dosage form that fits the desired route of administration (e.g., oral, injectable, or topical). The initial formulation is needed for testing on animals (nonclinical studies). Results from the nonclinical studies will be used to determine whether the product can be considered safe enough to be administered to humans.

Lixisenatide is designed based on a molecule, exendin-4, found in the saliva of a venomous lizard, the Gila monster. It is shown to regulate glucose metabolism and insulin secretion. The molecule resembles the human hormone, glucagon-like peptide-1 (GLP-1), which binds to the GLP-1 receptor and causes insulin production. While naturally occurring GLP-1 is only in circulation for a few minutes, synthetic forms can last for hours to days and have led to a new class of drugs known as GLP-1 receptor agonists. Lixisenatide is the sixth product to enter the market and has been modified with additional amino acid residues to resist degradation for a longer duration of activity (Fig. 3.2).

Rucaparib is a small molecule developed to treat advanced ovarian cancer with a specific genetic mutation in the *BRCA* gene. Because these cancer cells rely on an enzyme called poly(ADP-ribose)polymerase (PARP) for their survival, rucaparib was developed to inhibit the enzyme and selectively kill the cancer cells. Rucaparib is the second product in the class of drugs known as PARP inhibitors (Fig. 3.3).

[3]For example, 3-hydroxy-3-methyl-glutaryl-CoA (HMG-CoA) reductase is an enzyme involved in the synthesis of cholesterol in the human body. Although associations between fatty deposits in the blood artery and heart attacks were observed over 100 years ago, it took over 50 years to identify elevated cholesterol level as the main culprit for heart attack and another 40 years to develop a drug that can lower the blood cholesterol level. By inhibiting HMG-CoA reductase, lovastatin prevents the synthesis of cholesterol and lowers the level of cholesterol in the blood. The reason for the long timeline is that, typically, researchers need to synthesize and screen hundreds to thousands of compounds before they can select one that will be tested in a clinical trial. If the candidate is a therapeutic biologic, the process may involve screening cell culture broths or tissue extracts followed by the isolation and purification of the molecule of interest.

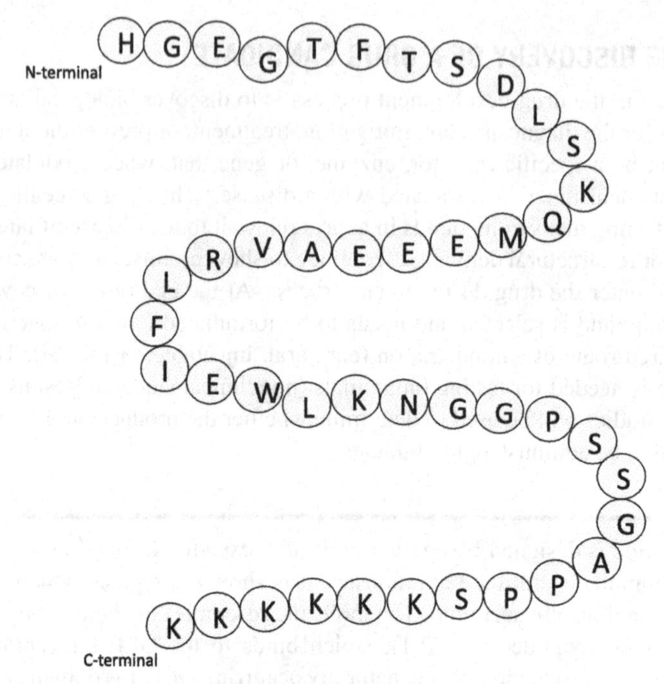

FIGURE 3.2 Structure of lixisenatide.

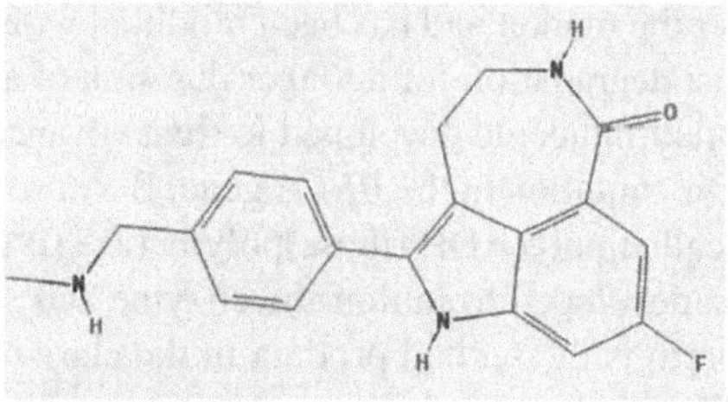

FIGURE 3.3 Structure of rucaparib.

3.4.2 THE NONCLINICAL PHASE OF DRUG DEVELOPMENT

A sponsor, at this stage of drug development, conducts in vitro laboratory studies and in vivo animal studies. The purpose of these studies is to determine whether the drug candidate shows drug activity associated with the therapeutic indication and whether it is reasonably safe to be used in a clinical trial. This is an important milestone in that the FDA and other regulatory agencies require demonstration of efficacy and safety in animal models before a compound can be studied in humans. These nonclinical studies can be categorized into pharmacology, pharmacokinetic, and toxicology studies. The information from these studies is used to design the initial clinical trial including estimating an initial safe starting dose and dose range for the first-in-human trials. In addition, the data is used to help identify parameters for clinical monitoring of potential adverse effects. The quality and reproducibility of the data obtained in nonclinical research are critical to ensuring the safety of human subjects. Efforts to harmonize requirements for registration of drugs internationally include nonclinical requirements. Most notable are the guidelines developed by the International Council for Harmonisation (ICH), which are widely adopted by the industry.

ICH guidelines for nonclinical safety testing requirements

- M3 Nonclinical safety studies
- S1A–S1C Carcinogenicity studies
- S2 Genotoxicity studies
- S3A–S3B Toxicokinetic and pharmacokinetics
- S4 Toxicity testing
- S5 Reproductive toxicology
- S6 Biotechnological products
- S7A–S7B Pharmacology studies
- S8 Immunotoxicology studies
- S9 Nonclinical evaluation for anticancer pharmaceuticals
- S10 Photosafety evaluation
- S11 Nonclinical safety testing

3.4.2.1 Pharmacology

One of the most important goals of nonclinical research is to determine whether a new drug will be safe and effective for the targeted disease. Sponsors conduct primary and secondary pharmacology studies to understand how the drug interacts with biological targets. The results of these studies help to assess whether the drug has the desired anticipated therapeutic effect. In addition, the results can also show whether there are any unexpected safety concerns with the drug.

Primary pharmacodynamic studies are performed to understand the biochemical, physiologic, and molecular effects of a drug on the body in relation to its desired therapeutic target. The objective of these studies is to provide evidence that the test compound demonstrates the desired pharmacological effect through laboratory assays and animal models.[4]

[4]Animal models are developed by identifying spontaneously occurring abnormal traits and selectively breeding the animals, so they exhibit the desired traits. More recently, with the development of modern biotechnology, applicants can also create genetically modified organisms. For example, mice with spontaneous mutation in the gene for leptin receptor are used to study diabetes and obesity (Bogdanov et al., 2014; Kobayashi et al., 2000). Genetically modified, mice-carrying mutant human superoxide dismutase 1 (SOD1) gene are used to study amyotrophic lateral sclerosis (Acevedo-Arozena et al., 2011). Sometimes, disease symptoms are induced using known toxins as in the case of 1-methyl-4-phenyl-1,2,3,6- tetrahydropyridine (MPTP)-induced mouse model of Parkinson's disease where the neurotoxin MPTP is used to induce the Parkinsonian syndrome (Jackson-Lewis & Przedborski, 2007). The primary pharmacology studies are considered exploratory, often conducted as part of discovery research. The regulatory requirements for these studies are not as stringent when compared to the toxicology studies described below. Unfortunately, data obtained from animal studies do not always translate well to human response and, therefore, proof of therapeutic efficacy of a compound requires testing in human clinical trials.

Lixisenatide was studied in diabetic mice, rats, and dogs to understand its potential therapeutic effect. In one study, lixisenatide was administered daily for 3 months to diabetic mice to examine its effects on disease progression. Data collected include fasting blood glucose, oral glucose tolerance, and HbA1c. The results showed that lixisenatide prevented the progressive development of diabetes in the diabetic mouse model. Other studies were conducted to measure how strongly lixisenatide binds to its biological target, GLP-1 receptor. In a study where lixisenatide and human GLP-1 were compared using mammalian cells transfected with human GLP-1 receptor, the results showed that lixisenatide has a binding affinity for the human GLP-1 receptor that is approximately four times greater than that of human GLP-1.

Rucaparib was studied using mouse xenograft models, where human cancer cells are grafted onto mice, to evaluate the drug's antitumor effects. In these studies, rucaparib administered orally for 28 days showed tumor growth inhibition in a dose-dependent manner. Enzyme inhibition studies were also conducted to examine rucaparib's ability to inhibit different types of enzymes. The results showed that the drug strongly and specifically inhibited PARP-1, PARP-2, and PARP-3 enzymes and only weakly inhibited others in the PARP family. Studies were also conducted in a panel of tumor cell lines to demonstrate that rucaparib causes increased cell death in tumor cell lines containing BRCA mutation compared to those without the mutation, thereby confirming the specificity of the drug's mechanism of action.

Secondary pharmacodynamics studies are performed to understand the "off-target" effects of a compound. They help predict potential adverse reactions to the compound and can be used to select a suitable lead candidate for clinical trials. Historically, products have been withdrawn after being on the market due to the appearance of unpredicted toxicity as in the case of terfenadine and cardiotoxicity. These secondary pharmacodynamics studies may overlap with—and can be incorporated into— the safety pharmacology studies described below.

Included in the secondary pharmacology studies for lixisenatide were receptor binding studies to assess how strongly the drug binds to other (off-target) receptors. When tested against 91 different receptors, lixisenatide showed little affinity to other receptors, indicating that it selectively binds to the target (GLP-1) receptor. The secondary pharmacology studies for rucaparib evaluated the potential of the drug to inhibit off-target receptors, enzymes, and channels using biochemical and cellular assays.

Information gathered from these studies are used to evaluate whether a drug exhibits an acceptable balance of therapeutic effects associated with the specific biologic targets and undesirable effects associated with off-target activities.

Sometimes, secondary pharmacology studies are used to explore other potential therapeutic targets. For example, cardioprotective effects of lixisenatide were studied in rats with induced heart failure (through ischemia/reperfusion injury) and mice genetically modified to exhibit atherosclerosis (apoE-knockout).

Safety pharmacology studies investigate the potential undesirable pharmacologic effects of a substance on vital organs or body systems that are critical to life. These include but are not limited to the central nervous, cardiovascular, and respiratory systems. Other physiological systems can be included as appropriate, based on a scientific rationale. These studies are designed to elucidate dose–response relationships between the compound and adverse effects so that the results can be compared to primary pharmacodynamics studies and help to inform the proposed therapeutic effect in humans.[5] The core battery of safety pharmacology studies should be completed before starting a phase 1 clinical trial. In some cases, safety pharmacology studies may not be necessary. Examples include well-characterized compounds that are locally applied and biotechnology-derived products targeting specific receptors.

> Safety pharmacology studies for lixisenatide examined behavioral effects in rats and mice and cardiovascular effects using hERG channel assay on cells and electrophysiology assay on rabbit Purkinje fibers. Other studies examined cardiovascular parameters like mean blood pressure in rats electrocardiogram changes in dogs, and gastric emptying in mice. Studies for rucaparib examined the drug's effects on hERG channel in cells. In addition, the studies examined the effects on blood pressure, heart rate, and cardiac rhythm in dogs, as well as the effects on nervous system function in rats.

3.4.2.2 Pharmacokinetics

For animal studies to be meaningful, it is important to understand the dose and exposure relationship. Pharmacokinetic studies are performed to understand the pharmacokinetic (PK) properties of a compound, including its absorption, distribution, metabolism, and excretion in test animals. This is done by measuring the maximum serum concentration (C_{max}), the time to maximum serum concentration (T_{max}), and the area under the curve (AUC) representing total drug exposure. The data from an animal pharmacokinetic study on the drug candidate are used to inform the in vivo toxicology program. The results are also used later in the development to compare human and animal metabolites to determine whether further studies should be conducted to characterize any specific metabolite(s).

3.4.2.3 Toxicology

The results of toxicology studies are used to predict toxicities and anticipate adverse events (AEs) in humans, as well as to select the clinical trial starting dose and provide initial information on the new drug's safety margin. The purpose of these studies is to examine the toxic effects of the compound including the types of toxicity,

[5]Examples of safety pharmacology studies to predict cardiovascular safety include in vitro and in vivo electrophysiology studies to detect the potential for delayed ventricular repolarization or QT interval prolongation.

affected target organs, dose dependency, AE relationship to drug level exposure, and potential reversibility of toxic effects. These studies must be conducted according to the Good Laboratory Practice (GLP) requirements to ensure the quality and integrity of the data. The toxicology studies should be designed to support the proposed clinical trials. Therefore, the sponsor should select the formulation used, the route of administration, the schedule, and the duration accordingly. Sponsors typically conduct single-dose toxicity and repeat-dose toxicity studies in two species (one rodent and one nonrodent mammal). Dose escalation studies are conducted to determine the maximum tolerated dose (MTD), which is sometimes referred to as the maximum feasible dose (MFD). The MFD is used to select the clinical trial dose range. Typically, a 50-fold exposure approach ($1/50^{th}$ MTD) is used in selecting the maximum dose for phase 2 and 3 clinical trials. Additionally, the sponsor may conduct other toxicology studies (including chronic repeat dose, genotoxicity, carcinogenicity, and reproductive) to support the clinical trials and the marketing application. The duration of repeated-dose toxicity studies depends on the maximum duration of the trial (see Table 3.1). Marketing applications require longer durations of repeated-dose toxicity studies (see Table 3.2).

Table 3.1 Duration of Repeated-Dose Toxicity Studies to Support the Conduct of Clinical Trials

Maximum Duration of Clinical Trials	Rodents	Nonrodents
Single dose	2 weeks	2 weeks
Up to 2 weeks	2 weeks	2 weeks
Between 2 weeks and 6 months	Same as clinical trial	Same as clinical trial
>6 months	6 months	9 months

Table 3.2 Duration of Repeated-Dose Toxicity Studies to Support a Marketing Application

Duration of Indicated Treatment	Rodents	Nonrodents
Up to 2 weeks	1 month	1 month
Between 2 weeks and 1 month	3 months	3 months
Between 1 month to 3 months	6 months	6 months
>3 months	6 months	9 months

Lixisenatide: Toxicology studies for lixisenatide development included a 6-month repeat-dose study in rats and a 12-month repeat-dose study in dogs. The results revealed reduced body weights and effects on male reproductive organs. Although these findings were observed at exposures that were over 4000-fold, the maximum anticipated human exposure, the company conducted a 6-month clinical trial to investigate the drug's effect on human sperm production. The study did not reveal any significant findings.

Genotoxicity studies for lixisenatide were conducted in bacterial cells (Ames test), mammalian cells (chromosome aberration test), and rodents (micronucleus test) all of which resulted in negative findings. Carcinogenicity studies performed in mice and rats over 2 years found an increase in thyroid cancer in animals treated with lixisenatide. However, these doses were over 1000-fold higher than expected clinical exposure, and the observed risk is similar to that seen with other GLP-1 agonists.

The results of reproductive and developmental toxicology studies showed that the drug caused skeletal malformation and the occurrence of rare visceral closure defects in rats and rabbits. As a result, the product labeling includes the language "Based on animal reproduction studies, there may be risks to the fetus from exposure to lixisenatide during pregnancy". In the European Union, lixisenatide has been approved with the tradename Lyxumia with the recommendation that insulin be used instead of lixisenatide during pregnancy.

Rucaparib: Toxicology requirements for anticancer drugs are more flexible due to the serious nature of the disease. Rucaparib was evaluated in repeat-dose studies in rats and dogs for up to 13 weeks. As expected for an oncology drug, rucaparib was found to be toxic to the embryo and caused DNA damage in cells. Carcinogenicity studies, fertility and early embryonic development studies, and prenatal and postnatal development studies were not conducted because they are not required for a drug that is intended to to treat patients with advanced cancer.

3.4.2.4 Estimation of the first dose in humans

The maximum recommended starting dose (MRSD) for first-in-human clinical trials for a new drug is typically determined using animal data. Specifically, the no observed adverse effect level (NOAEL), which is the highest dose level that does not produce a significant increase in adverse effect, is used to calculate the MRSD. The dose from the NOAEL is converted into a human equivalent dose by using a safety factor based on the body surface area. Usually, a conversion factor of 10 is used (i.e., the animal dose is converted into the human dose by dividing the animal dose by 10). Therefore, the doses for clinical trials would range from one-tenth of a NOAEL as the starting dose to 1/50th of MTD for the maximum dose. If there are no adverse findings in humans at the 1/50th dose, the clinical dose could be further escalated up to 1/10th of the MTD.

According to lixisenatide preclinical studies, NOAEL ranged from 200 to 2000 mcg/kg administered twice daily. The starting dose in clinical trials was 2 mcg. For a 70-kg person, the dose would be calculated as 0.03 mcg/kg, which is much lower than 1/10th of NOAEL even after applying the conversion factor of 10.

The NOAEL for rucaparib was determined to be 15 mg/kg according to the safety pharmacology study examining cardiovascular parameters in dogs. The starting dose in clinical trials was 12 mg, calculated to be 0.17 mg/kg for a 70-kg person, which is around 1/10th of the NOAEL after applying the conversion factor of 10.

3.4.3 THE CLINICAL PHASES OF DRUG DEVELOPMENT

In the United States, when a sponsor is ready to test the compound in humans, the sponsor will need to submit an Investigational New Drug (IND) application to the FDA. The application should contain the results of animal pharmacology and toxicology studies conducted to demonstrate that the product is reasonably safe for initial testing in humans. Manufacturing and control information should demonstrate that the product can be consistently manufactured under controlled conditions, has adequate quality, and is stable for the expected duration of the clinical development protocol(s). The application should include one or more clinical protocols and investigator documentation to demonstrate that the initial studies will not expose clinical trial subjects to unnecessary risks and that the investigator is qualified to oversee the study. At any point before the initial IND is submitted, the sponsor can contact the FDA and request a pre-IND consultation to seek the agency's guidance. The FDA has 30 calendar days to review the IND application before the sponsor can initiate the clinical trial. During this period, the sponsor may receive questions from the FDA or be subject to a "clinical hold" if the FDA has serious concerns. If there are no objections, the IND becomes in effect in 30 days and the clinical stage can begin. It is important to keep in mind that all US clinical trials must be entered into the clinicaltrials.gov database and before the protocol is initiated, it must be reviewed and approved by the appropriate IRB(s).

3.4.3.1 Study design

There are different study designs to consider when planning a clinical trial. The preferred study design in drug development is the prospective, randomized, double-blind, controlled type. In a randomized study, each study subject is randomly assigned to a group so that neither the subject nor the study personnel can anticipate the assignment. The group receiving the experimental drug is compared to the control group that received a placebo or, in some cases, a comparator. In a double-blind randomized trial, both the study personnel and the subject are "blind" to the assignment until after the study is completed. These controls allow an evaluation of the safety and efficacy of an investigational drug, while minimizing bias and confounding factors.[6] A randomized, controlled, double-blind design is considered to be the gold standard in clinical trials and one that is typically required by regulatory authorities as part of the marketing authorization for a new drug.

Unfortunately, most molecules fail during laboratory research or animal studies and never make it to the clinical trial stage. Similarly, for those that enter clinical trials, most do not meet the safety and efficacy hurdles and never make it to the market. During the earlier part of clinical development, researchers study the molecule in small, closely controlled trials to obtain information about its safety, tolerability, dosing, and pharmacologic effect(s). Because a new molecule poses risks that are yet unknown, clinical trials must proceed carefully in stages to minimize risks to the human study subjects.

3.4.3.2 Study phases

A new molecule undergoes clinical testing initially in phase 1 to assess safety and tolerability, then in phase 2 to examine efficacy and safety in the proposed indication, and finally in phase 3 to confirm both safety and efficacy in a larger population. The size of a clinical development program depends on many factors including the nature and incidence of the disease, the specific indication, and the identified regulatory pathway. For example, a drug that targets a broad indication such as diabetes will have a large development program with many clinical trials in different settings and will include a large number of patients. In addition, if a product has competitors on the market, the sponsor may design additional trials to differentiate their product from its competitors.

3.4.3.2.1 Phase 1 studies

During this first phase of clinical development, the investigational drug is administered to clinical trial subjects in a stepwise manner, starting with just a few subjects at the lowest possible dose as determined from animal studies. Once the safety of the first group of subjects is established, subsequent groups receive progressively higher doses in order to establish the MTD. These trials are usually performed in healthy volunteers in order to minimize the confounding effects of diseases and medications. In addition, phase 1 trials take place mostly in inpatient settings to provide maximum control over the conduct of the trial and to collect the necessary biological specimens to characterize the pharmacokinetic and pharmacodynamic profile of the molecule. The trial subjects are closely monitored for signs and symptoms of adverse events (AEs).

In certain situations, conducting phase 1 studies in healthy volunteers may not be feasible because of the toxic nature of the study drug. For example, anticancer drugs that are poisonous to normal cells would pose too great a risk to healthy volunteers and, as such, the choice of this particular subject population in these types of trials is unethical. Therefore, a first-in-human clinical trial to test an anticancer new drug usually is performed in subjects with advanced-stage cancer or those who are not responding to the current standard-of-care treatment for the particular cancer under study.

[6]Confounding factors may disguise or falsely demonstrate an apparent association between the treatment and an outcome when no association exists. In a clinical trial, this may happen, for example, when groups are compared that have a different distribution of a known prognostic factor.

Lixisenatide was studied in 19 phase 1 studies:

Nine single-dose studies in healthy subjects

Two single-dose studies in subjects with type II diabetes mellitus

Eight multiple-dose studies in healthy subjects

And five phase 1 studies with other formulations

Rucaparib, like most anticancer drugs, is considered too toxic to be administered to healthy subjects. Hence, the first-in-human study for rucaparib was conducted in cancer patients to evaluate the safety profile of the molecule and to determine the phase 2 doses.

3.4.3.2.2 Phase 2 studies

While the phase 1 of clinical development is primarily concerned with determining the safety and tolerability of the new drug, phase 2 focuses on exploring the effectiveness of the new drug in an indication where the molecule has demonstrated the desired effect in animals (such as increased survival or symptomatic improvement). These trials are performed in patients with the specific condition targeted by the investigational agent. In this early stage of development, very little is known about the molecule. The sponsor, typically, selects sites that have experience in conducting early phase trials and employs personnel with expertise in the therapeutic area of interest. A phase 2 clinical trial design can include several different groups or arms, each with a different dose of the study drug, a placebo, or a comparator drug. Although phase 2 focuses on detecting the therapeutic effect of the investigational drug and identifying the optimal dose range for the next phase of development, monitoring AEs continues to be a priority. The applicant has an interest in collecting safety information and developing an accurate understanding of the molecule's safety profile. If the data obtained in phase 1 and 2 clinical trials demonstrate that the trials have met their protocol endpoints, phase 3 clinical investigation(s) can proceed.

The two drugs, lixisenatide and rucaparib, followed very different clinical trial strategies.

Because lixisenatide targets a patient population that is very large, in a market that already has many competitors, the company carried out a large clinical development program with many trials to understand how the drug can be used in different settings. The phase 2 program had five trials comparing lixisenatide to placebo, lixisenatide to a drug that is already on the market, and different dosing paradigms.

In contrast, rucaparib targets a patient population that is very small, in a market that has only one competitor. The phase 2 program had three trials but, as an orphan indication, the number of patients in the clinical trials was small. Under "Accelerated Approval" and "Breakthrough Therapy" designations, the product received approval based on the results of phase 2 trials.

3.4.3.2.3 Phase 3 studies (pivotal trials)

The knowledge gathered from phase 1 and 2 trials is used to design one or more phase 3 trials to determine whether the new drug demonstrates adequate safety and effectiveness. These phase 3 clinical trials are usually very large with hundreds to thousands of subjects participating across the country or even across the globe. The reason for the large trial size is to ensure that the results obtained can be clearly and statistically attributed to the treatment. A phase 3 trial is usually randomized, double-blinded, and controlled with at least two arms (one for the study drug and one for the control). However, the study design may have more arms if, for example, the dosing scheme includes a placebo, and/or a comparator, and/or additional doses.

Data from 11 clinical trials were used to support that lixisenatide 20 μg administered subcutaneously once daily improves glycemic control in adult patients with type 2 diabetes mellitus, whose disease is not controlled by diet and exercise. The primary efficacy assessment was based on the change in hemoglobin A1c (HbA1c) from the baseline to the end of the treatment period. The broad phase 3 program included a monotherapy trial and those that added lixisenatide onto various existing therapies. Seventy percent of clinical subjects tested positive for lixisenatide antidrug antibodies (ADAs) after treatment with lixisenatide for 24 weeks or more.

3.4.4 THE CHEMISTRY, MANUFACTURING, AND CONTROLS COMPONENT OF DRUG DEVELOPMENT

During drug discovery and development, sponsors must collect information to ensure appropriate identification, quality, purity, and strength of the product. While organizations try to advance drug development as quickly as possible, they are all faced with questions about the product formulation, the manufacturing process, and the adequacy of analytical methods to control the product. Drug developers have competing concerns. If they over-engineer a process and invest too many resources early on, they will waste money and manpower if the drug then fails in clinical trials. On the other hand, if they disregard important aspects of pharmaceutical science during the early process development, a promising drug candidate may fail to have manufacturing, testing, and control processes that are viable and sustainable from a commercial standpoint. Regulators are interested in information about the quality of the drug substance and the drug product. This information is included in the quality section of a regulatory application, which is commonly called the Chemistry, Manufacturing, and Controls (CMC) section.

More broadly, the CMC describes the pharmaceutical development activities of the drug substance and the drug product. CMC activities are initiated early on in the process and are conducted in parallel as a development program advances from discovery into preclinical and clinical evaluation and into commercialization. The CMC body of information includes items such as:

- the chemical and physical characterization of the drug substance
- the qualitative and quantitative composition of the formulation
- the manufacturing facilities and processes
- the specifications, stability programs, and analytical methods

The purpose of CMC is to support the safety and effectiveness of the product by ensuring its quality. Sponsors and applicants have an obligation to ensure compliance with CMC regulations so that the manufacturing and testing processes do not pose a potentially unacceptable risk to the patient.

To be able to adequately evaluate the potential of a drug candidate, developers are required to make an upfront investment in formulation and product design. Practicing good science and building in quality from day 1 can help sponsors advance drug candidates efficiently. An effective regulatory CMC strategy is one that is appropriate for the phase of development and is driven by science. This approach helps to mitigate risk and minimize iterations from pilot formulations to the target product profile. It also helps to decrease potential delays as all CMC decisions are supported by a scientific justification.

CMC development activities usually start during discovery before the preclinical and clinical stages. Regulatory agencies will expect that the formulation used in phase 1 studies is suitable for administration in a study setting and meets safety expectations. During phase 2, the investigational product is typically required in larger quantities. Therefore, the manufacturing process is scaled-up. This step may include changes in the manufacturing process and the sponsor will need to evaluate whether those changes will have an impact on the characteristics, impurity profile, and stability of the product. Also, at this stage of the development program, the sponsor will have to plan how to address the results that have been collected so far in the regulatory filing application. For example, the development team must evaluate potential genotoxic impurities and the presence of polymorphs.

In some cases, the formulation might need to be modified before the product can enter a phase 3 trial. If this occurs, companies can conduct a bridging study to demonstrate pharmacokinetics equivalency across multiple formulations. By phase 3, the material used should be the proposed marketing formulation. Any formulation changes after this stage might need to be supported by costlier comparability or bioequivalence studies.

The two products in our example, lixisenatide and rucaparib, have different characteristics that influence how the products are formulated and presented.

Lixisenatide is a peptide, subject to digestion in the stomach and not amenable to oral formulation. It is, therefore, developed for subcutaneous delivery and presented as a pen injector, making it a combination product. The device portion of the product was reviewed by FDA's Center for Devices and Radiological Health (CDRH).

Rucaparib is a small molecule, developed as a tablet for oral administration. However, early stages of product development (preclinical and phase 1) were conducted using an intravenous formulation The oral formulation was introduced during phase 1 through a protocol amendment (Wilson et al., 2017).

3.5 WHAT ARE THE KEY STEPS OF A NEW DRUG APPROVAL PROCESS IN THE UNITED STATES?

If the results of clinical trials demonstrate that the investigational drug is safe and effective, the applicant will prepare and submit a New Drug Application (NDA) to the FDA's Center for Drug Evaluation and Research (CDER) to receive a formal approval to market the drug in the United States. Prior to submitting the application, the applicant can request a pre-NDA meeting with the FDA staff to discuss the content of the application and identify potential deficiencies. During this meeting, the applicant can also discuss the review designation options. A new drug can be reviewed either via a standard review process of 12 months (i.e., 60 days for the acceptance of the application and 10 months for its review) or via a priority review process of 8 months (i.e., 60 days for the acceptance of the application and 6 months for its review).

3.5.1 THE STANDARD NDA PROCESS

The NDA contains comprehensive information regarding the drug to demonstrate its quality, safety, and effectiveness and to determine whether the benefits of the drug outweigh its risks. The information contained in the NDA will be examined by the FDA reviewers to determine:

- Is the drug safe and effective for its proposed use?
- Do the benefits of the drug outweigh the risks?
- Is the drug's proposed labeling appropriate and complete?
- Are the methods used to manufacture the drug, and the process controls used to maintain the drug's quality, adequate to preserve the drug's identity, strength, quality, and purity?

If the FDA determines that a drug poses a high risk, it may require that the applicant develops a Risk Evaluation and Mitigation Strategies (REMS) program to ensure that the drug can be used safely. A REMS for an NDA may be:

- a simple medication guide; or
- a patient package insert to help patients avoid serious adverse events (SAEs); or
- a communication plan to inform healthcare providers about the risks of the drug; or
- a more extensive program, Elements to Assure Safe Use (ETASU), that may require specific training of healthcare providers, certification of pharmacies, or demonstration of laboratory test results for access to the drug.

All REMS need to be assessed at 18 months, 3 years, and 7 years after approval. The FDA may change the assessment frequency or eliminate the REMS program after 3 years if it determines that a REMS program is no longer necessary.

After an applicant submits an NDA, the FDA has 60 days to determine whether it will accept the file for review based on the initial assessment of the completeness of the application. The submission date refers to the date that the applicant first submits the application, whereas the filing date is the date that the FDA accepts the application. The FDA may issue a "refusal to file" letter if the application is determined to be deficient. In general, an applicant receives a "refusal to file" letter based on clear omissions of necessary information, or on severe inadequacies, that make the application incomplete and impede the start of a meaningful review. In this case, the applicant will need to resubmit the entire application. If the FDA files the NDA, the review period starts. During the review period, the applicant may receive requests from FDA for more information, to conduct alternative analyses, or to clarify information in the application. If the applicant would like to include additional data in an NDA that is under review, the FDA may extend the review timeline. During the review period, the FDA typically conducts inspections of the applicant's clinical sites (GCP), nonclinical sites (GLP), and manufacturing sites (GMP). The FDA can perform these inspections during any phase of product development. However, they are most likely to occur after the applicant submits the NDA. Another important activity that takes place during the NDA review period is the drug labeling negotiations. During the negotiations, the parties determine the content of the prescribing information (also called the package insert) that is intended to provide information to the healthcare professionals who will prescribe the drug. The information includes items such as the drug's indication, dosing information, warnings and precautions, adverse reactions, results of the clinical trials, and patient counseling information. This is a critical step in drug development because a product's labeling often drives its marketing and commercialization.

Following a preliminary review of the NDA, the FDA may convene an advisory committee meeting to seek independent opinions and recommendations from outside experts on specific questions regarding the safety and efficacy of the new drug. Advisory committee members may be asked to determine whether the data submitted in an NDA are adequate to support the approval of the drug; whether additional studies are needed; or whether changes should be made to a product's labeling. The recommendations made by the advisory committees are not legally binding but are considered by the FDA as it makes its final decision.

The FDA convened the Endocrine and Metabolism Drugs Advisory Committee to discuss lixisenatide as a component in a fixed-dose combination product along with insulin. The committee members raised concerns regarding ADAs that were seen in the lixisenatide clinical trials. This concern was translated into postmarketing requirements for the company to further study the immunogenicity of clinical trial samples.

At the conclusion of its review, the FDA issues either an approval letter allowing the applicant to market the drug or a complete response letter containing reasons why the FDA did not approve the drug. Within 48 hours after the FDA approves a drug, the agency posts the summary review documents on its web site along with the product labeling and the approval letter. Complete response letters are not made public.

The NDA for lixisenatide was designated as a standard review and approved approximately 1 year from the date of submission. As noted above, the advisory committee raised concerns regarding the appearance of ADAs that could potentially neutralize the function of lixisenatide (neutralizing antibodies) or cross-react with endogenous GLP-1 or glucagon to adversely impact regulation of glucose metabolism. Hence, the approval letter stipulates further investigation in this area. Specifically, the company is required to conduct a postmarketing study, using samples collected from the clinical trials, to determine the incidence of neutralizing antibodies and antilixisenatide antibodies that are cross-reactive with endogenous GLP-1 and glucagon peptides. The approval letter states that the FDA needs to review how the company will perform this study, including the assays that will be used. Furthermore, the letter specifies the timeline for the milestone deliverables.

The NDA for rucaparib was designated as priority review and approved approximately 6 months from the date of submission. Because it is approved under accelerated approval designation, the company is required to conduct further postmarketing clinical trials to demonstrate clinical benefit. The letter clearly states that the FDA may withdraw the approval if the postmarketing clinical trials fail to verify clinical benefit or are not conducted with due diligence. In this case, the drug was approved based on the surrogate endpoint of "objective response rate," which is defined as the proportion of patients with tumor size reduction of a predefined amount and for a minimum time period. The endpoint included in the postmarketing clinical trial is "overall survival," which is a universally accepted direct measure of clinical benefit. The approval letter specifies the timeline for milestone deliverables, including submissions of final protocol, interim report, and final report.

3.5.2 THE FDA-ACCELERATED PATHWAYS

An applicant with a drug that addresses an unmet medical need in the treatment of a serious or life-threatening condition can request the use of one of FDA's expedited programs: fast track, accelerated approval, priority review, and breakthrough therapy.

The fast track is a program that speeds time to approval by shortening the development and review period. To obtain a fast track designation, the applicant must submit a request to the FDA. The request can be initiated at any time and can be discussed with the FDA as early as the pre-IND meeting. The FDA will review the request and attempt to make a decision within 60 days. A drug that receives fast track designation is eligible for a rolling review. An applicant can submit completed sections of its NDA for review by FDA, rather than waiting until every section of the NDA is completed and submit it all at once before the review can begin. If the relevant criteria are met, drugs that receive a fast track designation are also eligible for accelerated approval and priority review.

The accelerated approval is a program that shortens the clinical study period. It allows drugs to be approved based on a *surrogate endpoint*.[7] Approval of a drug based on such endpoints is given on the condition that postmarketing clinical trials verify the anticipated clinical benefit. This speeds the availability of the drug to patients who need it. If the confirmatory trial shows that the drug provides a clinical benefit, then the FDA grants traditional approval for the drug.

The priority review is a program that speeds time to approval by prioritizing the agency review. Therefore, it shortens the review period. An applicant can request priority review. However, the FDA decides on the review designation. A priority review provides a shorter review time of 6 months compared with a standard review time of 10 months.

The breakthrough therapy is a program that aims to shorten the development period. It is designed to expedite the development and review of drugs where preliminary clinical evidence indicates that the drug may demonstrate substantial improvement over the available therapy. Breakthrough therapy designation is requested by the applicant. Ideally, a breakthrough therapy designation request should be received by FDA no later than the end-of-phase-2 meetings. A drug that receives breakthrough therapy designation is eligible for all fast track designation features. It also receives intensive guidance on an efficient drug development program, beginning as early as phase 1, and an organizational commitment from senior managers at FDA.

3.6 WHAT HAPPENS AFTER A DRUG PRODUCT RECEIVES REGULATORY APPROVAL?

In the United States, after the FDA approves an NDA, the applicant receives a license. The license holder's regulatory obligations continue after a drug is approved. For example, the license holder needs to monitor the drug's safety in order to understand how the drug performs outside of the confines of a clinical trial environment. This type of postmarketing surveillance is known as pharmacovigilance.[8] According to pharmacovigilance requirements, a license holder needs to submit reports on AEs in compliance with specific timelines. Sometimes, the new information collected through pharmacovigilance activities may trigger revisions to the product labeling, different types of communication to consumers and healthcare professionals, and

[7]A surrogate endpoint is a marker—can be a laboratory measurement, a radiographic image, or another measure—that is thought to predict clinical benefit, but is not itself a measure of clinical benefit. In oncology, for example, one might attempt to show that the treatment induces tumor shrinkage or delays tumor growth. Although these effects do not prove prolongation of survival, it is well known that patients with worsening levels of these markers have a greater risk for disease-related symptoms or death.

[8]Pharmacovigilance can be defined as the "science and activities that relate to the detection, assessment, understanding, and prevention of adverse effects or any other drug-related problem."

other strategies to minimize the risk. Other times, the new information may force a license holder to withdraw a drug from the market. This is the case if the drug's overall benefit/risk profile is no longer acceptable.[9] The FDA has also undertaken its own monitoring activities to look for "safety signals" of FDA-regulated medical products through the creation of a national electronic system, known as "The Sentinel System" (US Food and Drug Administration, 2008).

Some drug approvals come with the stipulation to continue clinical studies. These postapproval studies are known as phase 4 studies. The purpose of post-approval studies is to gather additional information about a product's safety or efficacy. The FDA makes a distinction between postmarketing requirements[10] and postmarketing commitments.[11] In both cases, depending on the study results, the license holder might need to amend the labeling for the product.

In addition, the license holder needs to evaluate changes to its approved products to assess their potential impact on the product's quality, safety and efficacy. All proposed changes should be evaluated and documented properly through a change-control process. Depending on the degree of impact, the regulatory requirements differ. Some changes can be implemented once the license holder documents that the change has been evaluated. Others can be initiated after notification to the FDA, whereas others require approval before implementation. Specifically, changes are reported as a supplement to the NDA, using one of four different reporting categories (US Food and Drug Administration, 2004) (see Table 3.3).

1. Annual Reportable Change: Used for a minor change that might have a minimal potential to affect the quality, safety, or efficacy of the drug product. These changes can be implemented *before filing*.
2. Changes Being Effected-0 (CBE-0) supplement: Used for a moderate change that might have a moderate potential to affect the quality, safety, or efficacy of the drug product. These changes can be implemented *immediately after filing*.
3. Changes Being Effected-30 (CBE-30) supplement: Used for a moderate change that might have a moderate potential to affect the quality, safety, or efficacy of the drug product. These changes can be implemented 30 days after filing (unless, within 30 days, the FDA raises objections).
4. Prior Approval Supplement (PAS): Used for a major change that might have a major potential to affect the quality, safety, or efficacy of the drug product. This type of change requires FDA approval prior to implementation.

[9]A well-known example is the one of Rofecoxib (Vioxx). Postmarketing data showed that patients taking the drug long term had an increased risk of heart attack. In 2004, Merck voluntarily withdrew the drug from all markets where Vioxx was approved and available.

[10]Studies that an applicant is required to conduct under one or more statutes or regulations are called postmarketing requirements.

[11]Studies that an applicant has agreed with the agency to conduct, but are not required by a statute or regulation, are called postmarketing commitments.

Table 3.3 Examples of Reporting Categories

Supplement Type	Example
Annual reportable change	A minor change in an existing code imprint for a dosage form (e.g. changing from a numeric to an alphanumeric code)
Minor change (CBE-0)	Elimination of in-process filtration performed as part of the manufacture of a terminally sterilized drug product
Moderate change (CBE-30)	Replacement of equipment with equipment of different design that does not affect the process methodology or process operating parameters
Major change (PAS)	Transfer of the manufacture of an aseptically processed sterile drug product to a newly constructed aseptic processing facility

3.7 HOW IS A NEW DRUG DEVELOPED AND APPROVED IN INTERNATIONAL MARKETS?

In developed countries, regulatory authorities administer laws and create regulations and guidelines to cover the development and commercialization of drugs. The systems and regulations governing pharmaceuticals, and the availability of and accessibility to medicines, vary in different regions of the world. The majority of the review and approval processes for prescription drugs is based on the criteria of quality, safety, and efficacy. However, regulators in different countries/regions act independently in implementing processes and systems to achieve these objectives and, as a result, national/regional laws, regulations, guidelines, and requirements vary. Historically, overlapping requirements in different countries/regions translated into duplication of work by the multinational companies that wanted to market their products globally. For example, a drug that was already approved and marketed in the United States still had to conduct a full clinical development in the local population, and sometimes, additional animal studies to meet the Japanese regulatory requirements. In the case of Europe, although the regulatory authorities accepted US preclinical and clinical data, the format of the European Marketing Authorization Application (MAA) was significantly different from the US NDA. These differences required a substantial amount of extra work to transform an NDA into an MAA or vice versa. This practice delayed patients' access to medicines. Over the last couple of decades, in an attempt to reduce healthcare costs globally as well as accelerate drug development, different regulatory agencies initiated efforts to work collaboratively. Most notably, in the 1990s, representatives from the regulatory authorities in the United States, Europe, and Japan (the three major markets at the time), in collaboration with industry associations, started a major international project: The International Conference on Harmonization of Technical Requirements for Registration of Pharmaceuticals for Human Use (ICH). One of ICH's initiatives was to harmonize some of the scientific and regulatory requirements in these three countries.

Since its inception, ICH has made significant strides toward harmonizing key aspects of drug development through the development of harmonized guidelines on safety, quality, and efficacy. For preclinical development, ICH guidelines provide recommendations on the amount of preclinical toxicity testing required to support clinical trials and market registration for a new drug. In the area of quality, ICH has produced harmonized guidelines on stability, specifications, and analytical methods to help define the quality of drug substances and drug products. The ICH efficacy guidelines help to minimize the duplication of clinical trials, harmonize the conduct of clinical trials to ensure the integrity of data, protect the safety of trial subjects, and facilitate reporting of clinical trial outcomes. Other important contributions by ICH include the common technical document (CTD). The CTD combines quality, safety, and efficacy information in a common format that is accepted by regulators not only in the ICH region, but also in other countries around the world (see Fig. 3.4).

Another important ICH guideline is the Medical Dictionary for Regulatory Activities, which standardized medical terminology and facilitates sharing of information.

In 2015, ICH changed its name to The International Council for Harmonisation and established itself as an independent legal entity under Swiss law. Since then, the

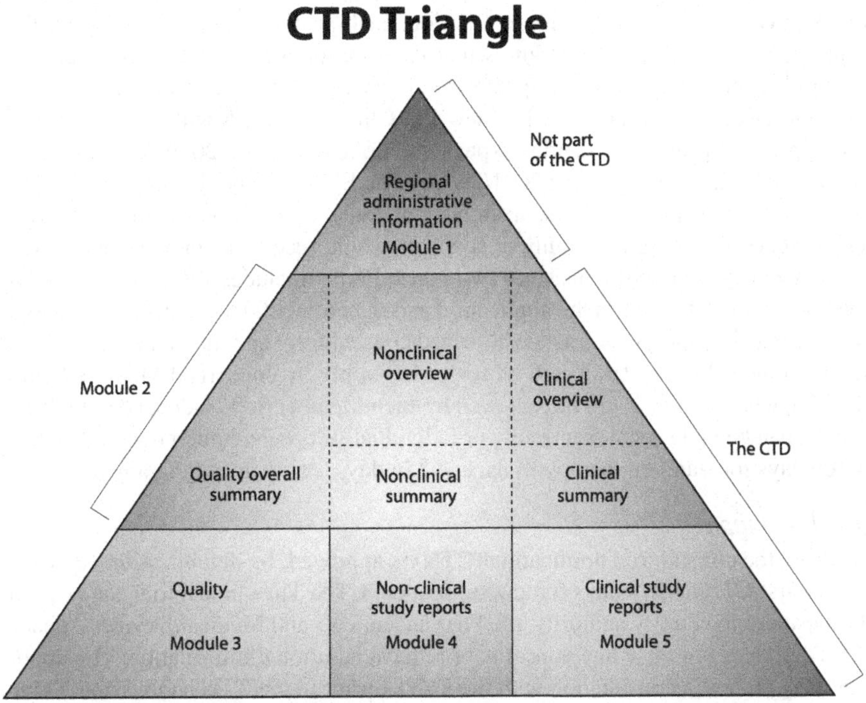

The CTD triangle. The common technical document is organized into five modules. Module 1 is region-specific and modules 2, 3, 4, and 5 are intended to be common for all regions.

FIGURE 3.4 Diagram of ICH CTD.

organization has made several changes and started new initiatives. Notably, the organization increased its international outreach and changed its governance structure.

3.7.1 KEY CONSIDERATIONS FOR STARTING CLINICAL TRIALS IN INTERNATIONAL MARKETS

Outside the United States, other countries have different regulatory requirements and processes for initiating clinical trials. For example, India does not allow first-in-human trials with products discovered abroad unless they satisfy specific requirements (e.g., the molecule targets a critical disease such as HIV/AIDS or some types of cancer). Examples of some local requirements for initiating clinical trials in a few selected countries are discussed below.

3.7.1.1 Europe

In Europe, the clinical trial application (CTA) process is not yet harmonized. Although the European Union (EU) Clinical Trials Directive 2001/20/EC strived to standardize the requirements for clinical trials in the EU, the process has resulted in longer timelines and increased costs. For a multinational trial to be implemented across the EU, separate CTAs need to be submitted to each Member State in which the sponsor wants to conduct a clinical trial. In addition, each protocol needs a separate CTA. Unlike the US IND, a CTA requires that the sponsor waits to receive the official approval from the National Competent Authority and the Ethics Committee before the sponsor can begin the trial. The new EU Clinical Trials Regulation (EU-CTR) was approved in April 2014 and is expected to go into effect in 2019. The new regulation will replace the directive 2001/20/EC. The EU-CTR should simplify the process by allowing sponsors to use a single application that can be submitted through a centralized EU portal. In addition, sponsors would receive a single decision. The harmonized dossier will consist of two parts. Part 1 includes the common study-specific documents such as the application and the protocol. Part 2 includes country/site-specific documents such as the informed consent form and the financial compensation to the subjects. Part 1 will be reviewed jointly by concerned Member States (CMSs), whereas part 2 will be reviewed by each Member State independently. Parts 1 and 2 can be reviewed concurrently or individually. The maximum review timeline is 106 days for standard therapy trials, and 156 days for advanced therapy trials.

3.7.1.2 Japan

In Japan, the clinical trial notification (CTN) is approved, by default, within 30 days for the first CTN and 14 days for any subsequent CTN. The clinical trial can begin if the Japanese regulatory authority, the Pharmaceuticals and Medical Devices Agency (PMDA), does not raise any concerns or request additional information. The applicant may request a consultation with the PMDA prior to the CTN submission. Some of the unique aspects of conducting clinical trials in Japan include J-GCP, which is slightly different from the ICH-GCP. The J-GCP emphasizes that the leader of the institution (clinical site head) needs to oversee the conduct of the study.

3.7.1.3 China

The length of time needed to start a clinical trial in China is significantly longer than it is for the United States. China's local clinical trial requirements are unique in that the size of the trials are stipulated according to the drug category (see Table 3.4).

The process begins with the submission of a CTA dossier to the China Food and Drug Administration (CFDA). The Center for Drug Evaluation (CDE), the technical review division, reviews the submission dossier and may request additional information. The CTA review timelines require a minimum of 175 working days, which makes it difficult for China to join multinational clinical trials. The applicant must submit investigational drug samples to the National Institute of Food and Drug Control (NIFDC) for testing and analysis and be ready to have its manufacturing sites inspected by the Center of Food and Drug Inspection (CFDI). The CDE completes the evaluation and submits its recommendation to the CFDA for or against the issuance of a Clinical Trial Permit (CTP). The entire process generally takes about 18 months. After obtaining a CTP, the applicant can only conduct studies at CFDA-certified sites.

3.7.2 THE REGULATORY APPROVAL PROCESS IN INTERNATIONAL MARKETS

Generally, when a multinational company considers bringing a new drug to the market, the company initially pursues a common regulatory strategy for the United States and Europe before considering other markets. The United States and European regions represent the largest markets. Typically, other countries such as Australia, Canada, New Zealand, South Africa, and Switzerland may follow. These countries can use a similar dossier as the one used for the US NDA, or the European MAA, with some adaptations based on local requirements. The adaptations will mostly relate to the ICH Module 1 or Module 3. Other countries, such as Japan, China, India, and South Korea, typically require that applicants include data from clinical trials conducted on subjects in their country's population in their marketing applications.

Table 3.4 Size of Clinical Trials by Product Category

Category	Study	Subjects
1 and 2	Phase1	20–30
	Phase 2	100
	Phase 3	300
	Phase 4	2000
3	Pharmacokinetic	100 pairs
4	Pharmacokinetic	100 pairs
5	Pharmacokinetic	100 pairs
	Bioequivalence	18–24
6	Bioequivalence	18–24

The timing for conducting this local clinical trial is critical. If the trial is initiated too late in the process, the applicant can experience long delays in bringing the new product to those markets. If, however, the trials are initiated too early, the applicant may find that the product has failed to meet its safety and efficacy endpoints in the main US/EU development program. In this case, the applicant will have spent unnecessary time and resources on redundant trials.

Countries in Latin America typically do not require that clinical trials be conducted on the country's population. However, some countries require that the applicant submits a Certificate of Pharmaceutical Product (CPP) to demonstrate that the drug is approved elsewhere, usually in the country of origin. This requirement represents additional delays in bringing a new product to the market especially if the applicant needs to submit the CPP with the marketing application.

3.7.2.1 Europe

To place a medicinal product on the market in the European Economic Area (EEA),[12] an applicant needs a marketing authorization (MA). There are four registration pathways for MA applications in the EU:

a) Centralized procedure
b) Mutual recognition procedure
c) Decentralized procedure
d) National procedure

Typically, applicants should choose a registration pathway at least 12–18 months before they plan to submit their application. Their choice should reflect the characteristics of the product and take into consideration the organization's business strategy.

3.7.2.1.1 European MA procedures

a) The Centralized Procedure

The centralized procedure allows applicants to submit a single application, go through a single evaluation, and receive a single MA, which provides direct access to the entire EEA. Products approved through the centralized procedure have a common Summary of Product Characteristics (SmPC, or SPC), which is the EU equivalent of the US Prescribing Information (USPI). The Centralized Procedure is *mandatory* for the following products:

- Biotechnological medicinal products (recombinant DNA technology, controlled gene expression, monoclonal antibodies)
- New active substances for specific therapeutic classes (new active substances indicated for the treatment of acquired immune deficiency syndrome (HIV), cancer, diabetes, neurodegenerative disorders, auto-immune diseases and other immune dysfunctions, and viral diseases)
- Orphan medicinal products
- Veterinary products used for performance enhancement
- Advanced therapies.

[12]The European Economic Area currently includes the EU Member States and the three EEA European Free Trade Association (EFTA) States (Iceland, Liechtenstein, and Norway).

The centralized procedure is *optional* for the following products:

- New active substances (medicinal products that constitute a significant therapeutic, scientific, or technical innovation or where the granting of a Community Authorization for the medicinal product is in the interests of patients at the community level)
- Immunological veterinary medicinal products for the treatment of animal diseases that are subject to community prophylactic measures.

For products in which the procedure is optional, the applicant needs to first submit an eligibility request to the Committee for Human Medicinal Products (CHMP). Four principal parties are involved in the centralized procedure: (1) the applicant, (2) the EMA secretariat, (3) the CHMP, and (4) the European Commission. The EMA appoints a project team leader for each application. This individual is responsible for all communication with the applicant. For the scientific assessment, the CHMP appoints a rapporteur and a co-rapporteur chosen from the CHMP members who represent each Member State. The choice is based on the therapeutic area expertise that is required for the review of the application. CHMP members are nominated by each Member State and serve on the committee for a renewable 3-year period. Fig. 3.5 shows the steps involved in obtaining an EU MA through the centralized procedure.

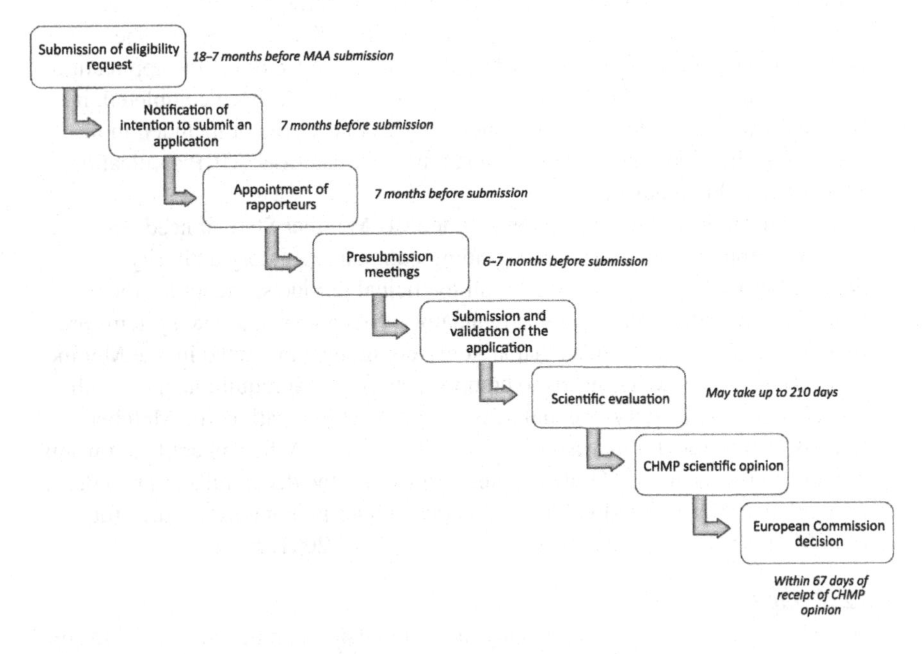

FIGURE 3.5 Overview of the centralized procedure review process.

b) The Mutual Recognition Procedure

The mutual recognition procedure (MRP) allows applicants to select a Reference Member State (RMS). This is a key difference between the MRP and the centralized procedure, where the rapporteur and co-rapporteur are assigned by the CHMP and not chosen by the applicant. The RMS conducts the initial assessment and grants the first approval as a national authorization. Once the applicant obtains national approval in the country serving as the RMS, it can initiate the MRP by asking the RMS to provide the authorization assessment to all countries identified as Concerned Member States (CMSs). All CMSs should recognize the first approval granted by the RMS within 90 days of the start of a mutual recognition procedure. Safeguard clauses allow the Member States to oppose mutual recognition if they consider that the product presents a potentially serious risk to public health. If an agreement between the Member States cannot be reached within 90 days, the points of disagreement are referred to the Coordination Group for Mutual Recognition and Decentralized Procedure (Human) (CMD(h)) for resolution. Then, if the Member States fail to reach consensus within the coordination group in 60 days, the issue is submitted to the EMA's scientific committee, CHMP, for arbitration. The opinion of the CHMP is then forwarded to the European Commission for the final decision.

c) The Decentralized Procedure

The decentralized procedure is available to applicants seeking approval for drug products that have not been previously authorized in any Member State and where it is not mandatory to use the centralized procedure. The applicant chooses one country to serve as the RMS and submits the application simultaneously also to the CMSs. Once the procedure has been completed, if the applicant wants additional MAs in countries that were not included in the original application, the applicant can submit a "repeat use" MRP application.

d) The National Procedure

If an applicant is seeking an MA in just one EU Member State, it needs to submit an application to the corresponding national regulatory authority. National procedures can be used for all medicinal products, except for those where the centralized procedure is mandatory. Nowadays, national systems are mostly used for drug products that are only going to be marketed in one Member State. The national MAA needs to comply with the local requirements, which often differ from country to country. For a national application, the Member State is responsible for assessing the risk–benefit ratio. When an applicant wants MAs from two or more Member States, it must use the decentralized procedure, rather than submitting individual applications under national procedures for each Member State (Council Directive 2001/83/EC, 2001, p. 78).

3.7.2.2 Japan

Until the late 1990s, the Japanese agency requested that all clinical data used to support an NDA in Japan be conducted on Japanese subjects in Japan. With the introduction of the ICH guideline ICH E5 "guideline on ethnic factors in the acceptability of foreign clinical data," the Japanese regulatory authority now accepts some non-Japanese clinical data to support an application for an MA. The process that allows the applicants to bring a new drug to the Japanese market consists of two key steps: (1) obtaining marketing approval and (2) getting the determination of the pricing and reimbursement rate. An MA application for a new drug, often called J-NDA, is prepared in CTD format and reviewed by the PMDA. Before submitting a J-NDA, the applicant should check all the documents in the application for completeness. It is not recommended to provide additional data or information after submission unless it is specifically requested by the regulatory authority. Also, once the review process has started, applicants are typically not allowed to have ongoing clinical trials with the study drug.

The J-NDA is submitted through the prefectural (regional) branch of the Ministry of Health, Labour, and Welfare (MHLW). Applicants are required to pay application fees according to the information posted on the PMDA web site. Once submitted, the J-NDA is transmitted to the PMDA and undergoes an administrative compliance check. During this step, the Japanese regulators verify the acceptance of the application for review and, in parallel, conduct a compliance verification in which the reliability of the data, GLP compliance, and GCP compliance are reviewed. If this first step is successful, the agency accepts the application and schedules the first meeting with the applicant to clarify potential issues before the application is fully reviewed. During the full in-depth review, the reviewers identify deficiencies and send the applicant questions and requests for clarifications. These requests stop the review clock for up to 12 months. The PMDA review team completes its review, and the review report is submitted to MHLW. The application is further reviewed by an additional expert committee, the Pharmaceutical Affairs and Food Sanitation Council (PAFSC), during which time additional questions may be identified. If necessary, an oral hearing, also called the Interview Review Meeting, is organized. This is typically a face-to-face meeting where the applicant meets the PMDA review team and the PAFSC experts. The applicant is required to send a briefing package to the PMDA two weeks prior to the hearing. The package should include an agenda, the list of experts that will attend the meeting, and responses to all questions raised by the PMDA and the PAFSC. After the oral hearing, the PMDA and the PAFSC officials hold an expert discussion behind closed doors. Following this discussion, the applicant is informed if there is an agreement on the approvability of the application. The PMDA then generates and submits a final report to the Evaluation and Licensing Division of the Pharmaceutical and Food Safety Bureau (PFSB), which is published and used as the basis for determining the drug classification for pricing.

3.7.2.3 China

China has a unique regulatory system. A drug can fall under one of the six categories as shown in Table 3.5 according to the CFDA.

Categories 1–5 require the submission of an NDA, and category 6 requires a generic drug application. There is a separate pathway for an imported drug that includes foreign new drugs. Import Licenses or Drug Product Certificates are valid

Table 3.5 Categories of Drugs

Category 1	A new drug which has never been marketed in any country
Category 2	A new drug preparation which changes the administration route of the marketed drug and has not been marketed anywhere in the world in this form
Category 3	A new drug which is already marketed outside of China, but has not been marketed in China
Category 4	A new drug which changes the acid or alkali (or metal) radical of a product marketed in China, but does not change the pharmacological effect
Category 5	A new drug preparation which has a different dosage form from drugs already marketed in China, but does not change the administration route
Category 6	Drugs that have already established national specifications in China (generics)

for 5 years. Renewal applications should be submitted 6 months prior to the product license expiration date along with post-marketing surveillance reports. The new drug registration application has four sections: (1) overview summary, (2) pharmaceutical sciences (CMC), (3) nonclinical pharmacological and toxicological research, and (4) clinical research. The application is submitted to the CFDA for initial examination and acceptance. Subsequently, the CDE performs the technical review during which the NIFDC tests drug samples. The NDA review and approval usually takes 7–15 months.

3.8 WHAT ARE THE KEY STEPS OF A GENERIC DRUG DEVELOPMENT AND APPROVAL PROCESS?

The development and the regulatory pathway that allows a generic drug to reach the US market is shorter and easier than the one for a new drug. The Drug Price Competition and Patent Term Restoration Act of 1984, also known as the Hatch-Waxman Act, amended the Federal Food, Drug, and Cosmetic Act, creating section 505(j). This section establishes an Abbreviated New Drug Approval (ANDA) pathway for generic drugs. This pathway allows a generic drug to rely on safety and effectiveness data that was performed by the brand drug company. As a result, developing a generic drug is much less expensive than developing a new drug.

3.8.1 THE ANDA PATHWAY

An ANDA submitted under 505(j) must have the same active ingredient(s), route of administration, dosage form, strength, and indication as the reference listed drug (RLD).[13] ANDA submissions must include information on manufacturing operations and controls, product testing, and labeling identical to the RLD. However, there may be differences in excipients and how the drug is supplied. The examples discussed below will help illustrate the development and approval pathways of a generic drug.

3.8.2 THE CONCEPT OF BIOEQUIVALENCE

A generic drug has the same indication, active ingredient(s), dosage form, strength, and route of administration as its corresponding RLD. The FDA requires applicants of a generic drug to provide evidence that would help establish *bioequivalency* between their drug and the reference drug. The agency uses two parameters when evaluating bioequivalence: one is the *rate* of absorption and the other is the *extent* of absorption. When a drug is administered, and serum concentrations are measured at various time points, the AUC can be calculated. The AUC represents the extent of absorption and is a cumulative measure of the drug's concentration over time. The rate of absorption is defined by the FDA as the highest or maximum concentration that can be detected, and it is referred to as C_{max}. When a brand drug is compared to a proposed generic equivalent, the active ingredient of the two drug products' rate and extent of absorption must only differ by −20%/+25% or have an overall bioequivalence of 80–125%.

At times, if scientifically justified, applicants may request and obtain a waiver from FDA for performing in vivo bioequivalence studies. In some cases, comparative dissolution data may be sufficient to show that the generic and the reference brand drug release drug at the same rate. Applicants generally submit "interim" dissolution data in their ANDAs. This is because, when they submit their application, they have not manufactured enough production-size batches to be able to establish the "final" dissolution values for the product release specifications. After the first three production-size batches are manufactured, the applicant can submit the dissolution data as a special supplement—Changes Being Effected (CBE). If there are no revisions proposed to the "interim" specifications, or if the final specifications are tighter than the "interim" specifications, a CBE is acceptable for the "final" dissolution specifications. If revisions need to be made to the interim specifications, or if the specifications are looser than the ones originally proposed in the "interim" dissolution specifications, these data should be submitted in the form of a Prior Approval Supplement (PAS). For some generic products, however, the FDA may require applicants to conduct comparative clinical studies.

[13]The FDA defines an RLD as "an approved drug product to which new generic versions are compared to show that they are bioequivalent."

Paroxetine hydrochloride extended-release tablets, 12.5 mg (base) and 25 mg (base), is one of Mylan's generic drugs. Paroxetine is an antidepressant in a group of drugs called selective serotonin reuptake inhibitors. In addition to depression, paroxetine can be prescribed to treat anxiety disorders, obsessive-compulsive disorder, and premenstrual dysphoric disorder.

Mylan's paroxetine hydrochloride extended-release tablets, 12.5 mg (base) and 25 mg (base), were determined to be bioequivalent and therapeutically equivalent to the RLD, Paxil CR extended-release tablets, 12.5 mg (base) and 25 mg (base), developed by GlaxoSmithKline (GSK).

Bioequivalence was shown by using dissolution profiles comparing the RLD to the generic drug.

The FDA required Mylan to use their proposed "interim" dissolution testing and specifications for the first three production-size batches. The FDA required dissolution testing be incorporated into the stability and quality control program using the same method proposed in Mylan's ANDA.

In addition to being bioequivalent, generic drugs are developed with the intent of being interchangeable. The rationale behind the program was that once a healthcare provider wrote a prescription, patients would be able to take either the brand drug or its generic form manufactured by any organization that had an approved application. In theory, the drugs would work the same way, and patients would not experience any additional adverse reaction from the generic drug as compared with the brand drug. In some instances, a healthcare provider needs to be involved. For example, this might be the case for drugs that have a narrow therapeutic window or drugs that have a complex release profile. There are certain drug classes that had received negative patient-perceived feedback when brand and generic drugs were switched. Some of these include: analgesics, antidepressants, hypnotics, and anticonvulsants. (Al-Jazairi, Bhareth, Eqtefan, & Al-Suwayeh, 2008; Jefferson, 2009). An example of a drug pair that was "interchanged" and received negative patient-perceived feedback for efficacy was Wellbutrin XL 300 mg (GlaxoSmithKline) and Budeprion XL 300 mg (Teva). The C_{max} and AUC confidence interval levels were met for the generic drug. However, patients experienced a lack of efficacy when using the generic, Budeprion XL 300 mg. Teva, the generic marketing firm, conducted a double-blinded clinical study to assess whether the brand drug and the generic were therapeutically equivalent. The results of the study showed that they were not equivalent, and Budeprion XL 300 mg was removed from the market (Boehm, Yao, Han, & Zheng, 2013).

In the United States, brand drug companies normally receive patent protection for 20 years from the date of filing. Patents and their expiration dates are listed in the FDA's publication titled "Approved Drug Products with Therapeutic Equivalence Evaluations" (the "Orange Book") for drug products.

Companies filing an ANDA must certify to one of the following in their submission with regard to the RLD:

- There have never been any patents on the drug (Paragraph I certification), or
- The RLD patent has expired (Paragraph II certification), or
- FDA should approve the generic version after the date the last patent expires (Paragraph III certification), or
- The patents are either unenforceable, invalid, or would not be infringed upon (Paragraph IV certification).

Generic companies filing an ANDA with a Paragraph IV certification must notify the brand drug company of the filing. The brand drug company has 45 days to file a patent infringement action against the generic company. After the filing, the FDA cannot approve the ANDA application until the generic company successfully defends the lawsuit or until 30 months have passed, whichever comes first. The first applicant to submit a substantially complete ANDA with a Paragraph IV certification is eligible for 180 days of generic drug market exclusivity. Generic companies with 180 days of market exclusivity can generate significant revenue because they are able to price their products slightly lower than the brand drug, impacting the brand drug's market share before the other generic companies enter the market.

> Pfizer's Viagra is an example of a drug approved by FDA that underwent a patent challenge by a generic company. Pfizer's original patent of sildenafil citrate for treatment of pulmonary arterial hypertension was due to expire in 2012. Teva's ANDA claimed that Pfizer's patent was invalid or there was no patent infringement. In 2010, Pfizer sued Teva over their ANDA claim that a second patent on the use of sildenafil citrate specifically to treat erectile dysfunction was unenforceable. The court upheld Pfizer's second patent, which is due to expire in 2019. This prevents Teva from receiving approval for a generic form of Viagra until October 2019.

3.8.3 DIFFERENCES BETWEEN THE "GENERICS" AND THE "505(B)(2)" PATHWAYS

With the passing of the Hatch-Waxman Act of 1984, applicants could submit an NDA using a third pathway, the 505(b)(2) regulatory pathway. This pathway allows applicants to develop a new drug product and refer to the safety and efficacy information on an active ingredient from a previously approved drug. For example, the new drug product might combine two previously approved active ingredients, or be a new dosage form, or use a new drug delivery mechanism. An organization may want to develop a new drug that uses well-known active ingredients from an approved product and target a new indication. In this case, the applicant would only be required to prove efficacy for the new indication. This can be done using a bridging study. The applicant could refer to information from the application of a previously approved reference drug. The application would include information about these well-known active ingredients and their previously reported safety and efficacy data.

The 505(b)(2) pathway is not restricted to reliance on published studies. The applicant may rely on previous FDA findings of safety and effectiveness of unpublished studies from another applicant's submission as long as it is no longer legally protected. Unlike generic drugs approved using the ANDA pathway, drugs approved using the 505(b)(2) pathway are not granted 180 days market exclusivity. The example below illustrates how an applicant used a 505(b)(2) application to modify the route of administration, dosage form, and dose schedule of a brand tablet.

An example of a drug using the 505(b)(2) pathway is Alkermes' Aripiprazole lauroxil (Aristada) extended-release injectable suspension.

Aripiprazole lauroxil (Aristada), a prodrug of aripiprazole, has been developed as an extended-release, intramuscular injection for once-monthly administration in the treatment of schizophrenia. A prodrug can be defined as a drug substance that needs to be converted into the pharmacologically active agent by metabolic or physicochemical transformation. After injection, aripiprazole lauroxil is converted into aripiprazole.

Use of Safety and Efficacy Data from RLD (Brand Drug)

Aripiprazole lauroxil was developed to meet the regulatory requirements under section 505(b)(2), using aripiprazole tablets, Abilify, as the RLD. The safety and efficacy data to support this formulation relied, in part, on the data from studies conducted with aripiprazole tablets (Abilify).

Clinical Trials Conducted in Support of the 505(b)(2) Submission

Additional phase 1 and phase 3 clinical trials were required for the new intramuscular method of administration and the once-monthly formulation. The first four phase 1 trials were dose escalation studies administered in 2 different intramuscular sites: gluteal and deltoid muscles. One phase 3 study for efficacy of 2 doses was performed using the gluteal muscle administration site. Two additional extended-open label phase 3 studies (extensions of the first phase 3 study) were performed to obtain additional safety information.

Justification for FDA Approval

The FDA was able to approve Aristada based on previous safety and efficacy data from the oral aripiprazole tablets (Abilify), along with additional pharmacokinetic data submitted by Alkermes. The studies conducted by Alkermes provided evidence that the serum concentrations of oral aripiprazole given daily at approved doses were similar to serum concentrations of aripiprazole lauroxil given monthly at the studied doses.

Exclusivity

Aripiprazole lauroxil is an ester of *N*-hydroxymethyl aripiprazole and not an ester of aripiprazole. As such, aripiprazole lauroxil was determined to be an active moiety not previously approved by FDA and designated as a "New Chemical Entity." Therefore, the drug was given 5 years of market exclusivity.

Postmarket Reporting

There were no postmarketing requirements imposed on Alkermes for Aristada.

3.9 WHAT ARE THE KEY STEPS OF AN OVER-THE-COUNTER (OTC) DRUG DEVELOPMENT AND APPROVAL PROCESS?

A drug that can be purchased without a prescription is referred to as an OTC drug. It can be used safely for the conditions indicated without the need for a healthcare provider's supervision. Many products that consumers buy in the stores contain OTC drugs. Examples include sunscreen products, face creams, toothpaste, nail hardeners, and hand soaps. If a product makes a disease claim like "antimicrobial," or a structure/function claim like "builds collagen," the product would be classified as a drug. There are different pathways to the commercialization of OTC products: the NDA process (described earlier in this chapter), the monograph process, and the prescription to OTC switch. We will describe the two latter processes below.

3.9.1 THE MONOGRAPH PROCESS

A new drug can avoid extensive regulation if it is "generally recognized as safe and effective" (GRAS/GRAE) by the FDA. To be designated as GRAS/GRAE, a drug must be evaluated by qualified experts and determined to be safe and effective for use under the conditions prescribed, recommended or suggested in the labeling (Federal Food, Drug, Cosmetic Act of 1938). Such general recognition must be based on adequate published data demonstrating the drug's safety and efficacy. In addition, the drug should have been used for an adequate period under the labeled conditions. Effectiveness requirements for an OTC drug product must be supported by the same level of evidence as required for an NDA. This includes adequate and well-controlled studies showing efficacy of the drug. It is difficult to qualify a drug as GRAS/GRAE. In 1972, the FDA began the OTC review program of active ingredients in each monograph category. This is a 3-phase process where each phase must be published in the Federal Register.

During phase 1, advisory panels evaluate the active ingredients to assess their safety and effectiveness in self-use. Additionally, the panels review claims and recommend labeling, including therapeutic indications, dosage, and side effect warnings. At the end of this phase, the panel issues its report as an "advanced notice of proposed rulemaking" (ANPR) in the Federal Register and places the active ingredient into one of three categories:

- Category I: generally recognized as safe and effective for the claimed therapeutic indication
- Category II: not generally recognized as safe and effective or unacceptable indications
- Category III: insufficient data available to permit final classification

Phase 2 of the review is undertaken by the FDA, which reviews the comments received from the first Federal Register publication as well as the advisory panel report. Once the FDA finishes its review, the agency publishes a tentative final monograph in the Federal Register and solicits public's review and comment. This process is known as "notice of proposed rulemaking".

The final review, phase 3 will include FDA's review of all public comments and any new, relevant data obtained on the ingredient. The agency will then publish a rule in the Federal Register establishing a Final Monograph. These monographs stipulate the conditions for safety and effectiveness, including labeling claims, dosage strengths, and other conditions. The effective date of the Final Monograph is usually 1 year after publication in the Federal Register. OTC monographs can be reopened and updated or amended by publishing the intent in the Federal Register. OTC drug monographs may need to be updated to add, change, or remove ingredients, labeling, or other pertinent information. This process is known as a "proposed monograph amendment" and will be open for public comment. The review process will proceed as explained above.

3.9.2 THE PROCESS FOR RX-TO-OTC SWITCH

Rx-to-OTC switches are becoming common in today's drug market. The subjects of Rx-to-OTC switches are drugs that once required a prescription and subsequently were switched to an OTC drug. There are three ways to gain approval from the FDA to switch:

1. A drug manufacturer can supplement the drug's existing NDA,
2. The FDA can create or amend an OTC monograph,
3. The drug manufacturer can petition the FDA.

The most common way to gain FDA approval for an Rx-to-OTC switch is by supplementing the drug's NDA. This method allows the manufacturer to obtain market exclusivity for the switched product. If the Rx-to-OTC switch supplement involves a change to the drug that is supported by new clinical investigations that are conducted or sponsored by the applicant and are essential to the supplement's approval, the applicant will gain 3 additional years of marketing exclusivity for the OTC product (Piña, 2014).

REFERENCES

Acevedo-Arozena, A., Kalmar, B., Essa, S., Ricketts, T., Joyce, P., Kent, R., et al. (2011). A comprehensive assessment of the SOD1G93A low-copy transgenic mouse, which models human amyotrophic lateral sclerosis. *Disease Models & Mechanisms*, *4*(5), 686–700.

Al-Jazairi, A., Bhareth, S., Eqtefan, I., & Al-Suwayeh, S. (2008). Brand and generic medications: are they interchangeable? *Annals of Saudi medicine*, *28*(1), 33.

Boehm, G., Yao, L., Han, L., & Zheng, Q. (2013). Development of the generic drug industry in the US after the Hatch-Waxman Act of 1984. *Acta Pharmaceutica Sinica B*, *3*(5), 297–311.

Bogdanov, P., Corraliza, L. A., Villena, J., Carvalho, A., Garcia-Arumí, J., Ramos, D., et al. (2014). The db/db Mouse: A useful model for the study of diabetic retinal neurodegeneration. *PLoS One*, *9*(5), .

Council Directive 2001/83/EC of 6 November 2001 on the Community code relating to medicinal products for human use. (2001). *Official Journal of the European Union*, L311, 77–78.

DiMasi, J., Grabowski, H., & Hansen, R. (2016). Innovation in the pharmaceutical industry: New estimates of R&D costs. *Journal Of Health Economics*, *47*, 20–33.

Federal Food, Drug, and Cosmetic Act of 1938, 21 U.S.C. § 321.

Jackson-Lewis, V., & Przedborski, S. (2007). Protocol for the MPTP mouse model of Parkinson's disease. *Nature Protocols*, *2*(1), 141–151.

Jefferson, J. W. (2009). Antidepressants: brand name or generic? *Psychiatric Times*, *26*(5), 26–31.

Kobayashi, K., Forte, T., Taniguchi, S., Ishida, B., Oka, K., & Chan, L. (2000). The db/db mouse, a model for diabetic dyslipidemia: Molecular characterization and effects of western diet feeding. *Metabolism*, *49*(1), 22–31.

Piña, K. (2014). *A practical guide to FDA's food and drug law and regulation*. Washington, DC: The Food and Drug Law Institute.

US Food and Drug Administration. (2004). Guidance for industry: changes to an approved NDA or ANDA.

U.S. Food and Drug Administration. (2008). *Sentinelinitiative.org*. Retrieved from https://www.sentinelinitiative.org.

Wilson, R., Evans, T., Middleton, M., Molife, L., Spicer, J., Dieras, V., et al. (2017). A phase I study of intravenous and oral rucaparib in combination with chemotherapy in patients with advanced solid tumours. *British Journal Of Cancer*, *116*(7), 884–892.

CHAPTER

4 Biologics

Mary Ellen Cosenza

MEC Regulatory & Toxicology Consulting, LLC, Moorpark, CA, United States

"Childhood vaccines are one of the great triumphs of modern medicine. Indeed, parents whose children are vaccinated no longer have to worry about their child's death or disability from whooping cough, polio, diphtheria, hepatitis, or a host of other infections."

– Ezekiel Emanuel

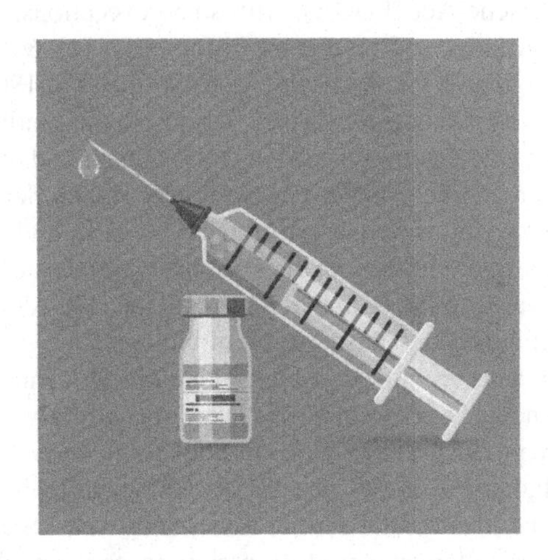

An Overview of FDA Regulated Products. http://dx.doi.org/10.1016/B978-0-12-811155-0.00004-1

CHAPTER OBJECTIVES

After reading this chapter, the reader will be able to:

- describe the differences between the regulatory pathways for biologics versus traditional small molecules,
- identify the differences between biologics and traditional small-molecule drugs,
- recognize that "biologics" is a broad category of medical products, and
- describe the regulations that impact the development of biologics.

4.1 WHAT IS A BIOLOGIC?

Once you determine that the Food and Drug Administration (FDA) regulates your product, you will need to determine its classification. How do you know if it is a "biologic"? Chapter 3 provided an overview of the three major categories of drugs—new drugs, generic drugs, and over-the-counter drugs—and the different regulatory pathways that are associated with bringing these products to the market. This chapter will focus on those products that are classified as "biologics." In the United States, biologics are largely governed by the Public Health Service Act (PHSA), not just the Food, Drug, and Cosmetic Act (FDCA), with some exceptions. This difference is derived from their legislative and regulatory history (Korwek & Druckman, 2015). Biologics were first regulated via the Biologics Control Act of 1902, which focused on the "safety, purity, and potency of vaccines, serums, toxins, antitoxins, and similar products." This Act sets the regulatory pathway for biologics and their manufacturing facilities and became recodified in 1944 as the Public Health Service Act (PHSA). Biologics were regulated by different government services and agencies over the years and did not move to the FDA from the National Institutes of Health (NIH) until 1972. The goal then was to apply both the PHSA and the FDCA (safe, effective, and not misbranded or adulterated) to the approval of biologics.

What medical products are now considered "biologics"? This category of products has grown tremendously in recent years. There are currently more than 250 biotechnology healthcare products and vaccines available commercially across the globe (www.bio.org). The Public Health Service Act (PHSA) states that "the term 'biologic product' means a virus, therapeutic serum, toxin, antitoxin, vaccine, blood, blood component or derivative, allergenic product, protein (except any chemically synthesized polypeptide), or analogous product, or arsphenamine or derivative of arsphenamine (or any other trivalent organic arsenic compound), applicable to the prevention, treatment or cure of a disease or condition of human beings" (42 United States Code, 262(i), n.d.). They are medical products that can be made of sugars, proteins, or nucleic acids or complex combinations of these substances, or may be living entities such as cells and tissues. These would include cell and gene therapies, therapeutic viruses, and CAR-T therapies. Other common terms are biopharmaceuticals, biologics, and therapeutic proteins. It is clearly no longer just blood products and vaccines.

The primary difference between biopharmaceuticals and traditional pharmaceuticals is the method of manufacture and processing. Biologics are manufactured in living organisms such as bacteria, yeast, and mammalian cells, whereas traditional "small-molecule" drugs are manufactured through a series of chemical synthesis. (There is a grey area with chemically synthesized peptides.) Many biotechnology products are produced using recombinant DNA technology.

There are several subsets of biologics and we will focus on the FDA designations, but it is worth mentioning some other jurisdictional definitions. The International Council for Harmonization (ICH) first defined the term biologics in the ICH S6 guidance: Preclinical Safety Evaluation of Biotechnology-Derived Pharmaceuticals. In that guidance biopharmaceuticals are defined as: Products derived from characterized cells including bacteria, yeast, insect, plant, and mammalian cells. Includes proteins, peptides, their derivatives, or products of which they are components. Examples include: Cytokines, proteins, growth factors, fusion proteins, enzymes, receptors, hormones, and monoclonal antibodies (International Council for Harmonisation, 1997c (ICH S6)).

The EMA generally follows the ICH definition of biologics, but it is worth noting that they have a Committee for Advanced Therapies (CAT). This Committee is responsible for assessing the safety and efficacy of advanced therapy medicinal products (ATMPs). These are defined as medicines for human use that are based on genes, cells, or tissue engineering (The European Parliament and The Council of the European Union, 2007).

Like drugs, some biologics are intended to treat diseases and medical conditions. Other biologics are used to prevent or diagnose diseases. Today, examples of biological products include: Vaccines, blood, and blood products for transfusion and/or manufacturing into other products, allergenic extracts, which are used for both diagnosis and treatment (e.g. allergy shots), human cells and tissues used for transplantation (e.g. tendons, ligaments, and bone), gene therapies, cellular therapies, and tests to screen potential blood donors for infectious agents such as HIV (U.S. Food and Drug Administration, 2017c).

4.2 WHICH CENTER AT THE FDA REGULATES BIOLOGICS? CBER OR CDER?

In the United States, the FDA is divided into several Offices (Office of the Commissioner, Office of Foods and Veterinary Medicine, Office of Medical Products and Tobacco) and then within each of these Offices there are Centers. These Centers focus on specific categories of products, for example, the Center for Food Safety and Applied Nutrition (CFSAN) or the Center for Tobacco Products. Medical products are divided into three main Centers: the Center for Drug Evaluation and Research (CDER), the Center for Biologic Evaluation and Research (CBER), and the Center for Devices and Radiological Health (CDRH). "Biologics" may be regulated by the Center for Drug Evaluation and Research (e.g. therapeutic proteins and monoclonal antibodies), the Center for Biologic Evaluation and Research (e.g. vaccines, blood products, cell and gene therapies), or even the Center for Devices and Radiological Health (e.g. combination products).

Before 2003, most "biologics" were regulated by CBER (The Center for Biologics Evaluation and Research), and traditional pharmaceutical small molecules were regulated by CDER (The Center for Drug Evaluation and Research). The exceptions were hormones, oligonucleotides, and chemically synthesized peptides. In the early 2000s, there was a reorganization that led to the transfer of therapeutic proteins (well-characterized proteins), including monoclonal antibodies, to CDER. This move was designed to take advantage of the therapeutic expertise in CDER as biopharmaceuticals became more prevalent and mainstream. CDER now regulates several categories of biological products including replacement proteins, monoclonal antibodies, including those modified by conjugation to small molecules drugs (antibody–drug conjugates or ADCs) or parts of antibodies (Fc fusions) and cytokines, growth factors, enzymes such as thrombolytics and immunomodulators (see Table 4.1). CBER is divided into Offices based on modalities. For example, the Office of Blood Research and Review, the Office of Vaccines Research and Review, and the Office of Tissue and Advanced Therapies, which review cell and gene therapies. CDER is further divided into Offices largely based on therapeutic areas (e.g. Office of Hematology and Oncology and the Office of Antimicrobial Products).

The licensing process for biologics still follows PHSA, even for products no longer regulated by CBER. The following table outlines the responsibilities of CBER and CDER with regards to biologics.

4.3 WHAT IS FDA'S ROLE REGARDING BIOLOGICAL PRODUCTS?

FDA's regulatory authority for the approval of biologics resides in the Public Health Service Act (PHS Act) and the Federal Food, Drug, and Cosmetic Act (FD&C Act). CBER, under the FD&C Act's Medical Device Amendments of 1976, regulates

Table 4.1 Regulation of Biotechnology Products

Products Regulated by CDER	Previously Regulated by CBER, now by CDER	Products Still Regulated by CBER
Hormones	Proteins and modified proteins	Gene therapies
Chemically synthesized peptides	Cytokines	Cellular therapies
Oligonucleotides	Growth factors	Engineered tissue products
	Ligands and receptors	Vaccines
	Antibodies	Blood and blood products
		Antitoxins

some medical devices used to produce biologics. A combination of Centers may be involved in the regulation of combination products, including CDRH. There is more discussion about combination products in Chapter 6.

Some of the direct responsibilities of FDA in regulating biologics include the review and approval of products for marketing; inspections of manufacturing plants before product approval and on a regular basis thereafter; review and approval of new indications for already approved products; and monitors the safety of biological products after they are marketed. As with drugs, the FDA also provides the public with information to promote the safe and appropriate use of biological products.

Specifics of the Public Health Service Act also give the FDA authority to approve biological products and immediately suspend licenses where there exists a danger to public health. It grants the FDA the power to procure products in the event of shortages and critical public health needs and to enforce regulations to prevent the introduction or spread of communicable diseases (42 United States Code, 262(i), n.d.; U.S. Food and Drug Administration, 2009).

Prior to 1997, biologics manufacturers had to file both a PLA (Product License Application) and an Establishment License Application (ELA) for approval to market their biologic product. In 1997, the Food and Drug Administration Modernization Act (FDAMA) attempted to harmonize the biologics and drug regulations. One change was consolidating the ELA and PLA into a single biologics license application (BLA).

A BLA includes manufacturing information to demonstrate that the company can properly and consistently manufacture the product and includes data and information regarding the biological product (animal and human studies). Data must demonstrate the product is safe and effective for its proposed use and can be manufactured consistently, meeting quality standards. The BLA also includes proposed labeling for uses for which it has been shown to be effective, possible risks, and how to use it. A BLA is very similar in approach to an NDA but generally includes more information on the manufacturing facility. As noted above, the Public Service Act authorized the issuance of a facilities license (dual licensing). This allows for rapid withdrawal of product from the market if a hazard arises.

4.4 HOW DO BIOLOGICS DIFFER FROM CONVENTIONAL DRUGS?

Most drugs consist of pure chemical substances and their structures are known. Most biologics, however, are complex mixtures that are not easily identified or characterized (see Table 4.2). Biological products differ from conventional drugs in that they tend to be heat-sensitive and susceptible to microbial contamination. This requires sterile processes to be applied from initial manufacturing steps. There are several preclinical parameters that differ as well. They can be immunogenic, are not metabolized, and their toxicity is often derived from their pharmacology.

Table 4.2 Differences Between Biologics and Pharmaceuticals (Cosenza, 2010)

Parameter	Small-Molecule Pharmaceuticals	Biotechnology-Derived Pharmaceuticals
Size	<500 Da	>1 kDa (macromolecules)
Immunogenicity	Nonimmunogenic	Potential for immunogenicity
Metabolism	Metabolized	Degraded
Frequency of dosing	Daily	Variable
Toxicity	Often structure based	Exaggerated pharmacology
Half-life	Short	Long
Route of administration	Usually oral, some IV	Parenteral
Species specificity	Active in most species	Species specific
Synthesis	Organic chemistry (synthesized)	Genetic engineering (derived from living material)
Structure	Well defined	Often not fully characterized

4.4.1 MANUFACTURING, QUALITY, AND CONTROLS

Biologics are a category of medicinal products that include a large variety of modalities. Some are clearly defined, such as blood product and vaccines, but others can be complex mixtures (cell therapies). Because biologics are derived from living matter, they can be variable in nature and are often comprised of complex mixtures. They are not easy to define or characterize and can be susceptible to contamination. Even replacement proteins such as erythropoietin and filgrastim (G-CSF) have been modified. Protein science groups have found ways to modify proteins through posttranslational modifications such as glycosylation and pegylation or create new constructs such as Fc fusions and bi-specifics. Because the products are derived from living organisms, their "manufacturing" is often extremely complex and requires additional controls. Many of the impurities are host proteins, nucleic acids, and viruses. The manufacturing must be conducted aseptically. An overarching theme often heard is that the "product is the process" which speaks to how small process changes can change the final product as well.

The unique characteristics of biologics contribute to the complication and challenges of manufacturing. In addition to the building blocks, amino acids and proteins have complex higher-order structure that are pivotal to ensuring the correct activity of these molecules. There are a series of ICH guidelines (see Table 4.3) that address many of the specific chemistry, manufacturing, and controls (CMC) issues for biologics (International Council for Harmonisation, 1997a, 1997b).

These guidelines help address some of the unique biologic product challenges such as screening of starting materials, purification steps, virus removal steps, viral inactivation steps, and validation-viral clearance studies.

Unique Drug Substance requirements for biologics include details on the cell expression system, for example, characterization of the master cell bank (MCB) and

Table 4.3 ICH Quality Documents for Biologics

Q5 A	Viral Safety Evaluation
Q5 B	Genetic stability (analysis of expression construct)
Q5 C	Product stability testing
Q5 D	Derivation and characterization of cell substrates
Q5 E	Comparability due to changes in manufacturing

working cell banks, cell growth and harvesting, and protein purification and downstream processing. For the Drug Product information on the formulation, fill and finish is key. Biologics must be processed under tightly controlled conditions/controls throughout production to consistently produce a safe, pure, and potent product and preclude the introduction of environmental contamination.

The PHS Act emphasizes manufacturing control and adherence to processes. Novel biological products are on the cutting edge of medical science and research. New biologic constructs often require new and novel manufacturing methods and quality control processes. Manufacturing processes for cell and tissue therapies, autologous or not, often require close collaboration with regulatory agencies. Newer therapies, like those using CAR-T cells, need to develop their processes, specifications, and controls as these therapies are being developed.

Manufacturers of biologics must comply with the appropriate laws and regulations relevant to their biologics license and identifying any changes needed to help ensure product quality. All GMP regulations must be adhered to. Certain types of manufacturing and production problems must be reported to FDA's Biological Product Deviation Reporting System (BPDR). There are established timeframes for reporting and corrective actions. There may be the need to recall or stop the manufacture of a product if a significant problem is detected.

4.4.2 PRECLINICAL DEVELOPMENT

The safety evaluation of biotechnology products has been an evolving process in the last three decades. ICH S6 guideline (Preclinical Safety Evaluation of Biotechnology-Derived Pharmaceuticals) help give guidance to preclinical scientists (both in industry and government) on how to approach these development plans. In the future, the basic questions for these products will remain the same, but the challenges will be greater as the molecules become more complicated. Global strategies for the development of biopharmaceuticals are generally be more science-driven and problem-focused. If industry scientists want the reviewers at regulatory agencies to stick to the scientific principles addressed in ICH S6, then the safety assessment industry scientist must be able to use science (e.g. target liability) to justify development plans as well. Defaulting to ICH S6 without scientific justification is as imprudent as conducting traditional small-molecule pharmaceutical studies on biopharmaceuticals inappropriately.

The overarching goal of any preclinical program is to evaluate the potential effects of these molecules and to provide guidance to the clinicians. Safety assessment approaches should have a strong scientific rationale and use the appropriate animal species judiciously. As with all pharmaceutical products, preclinical safety studies are expected to be conducted in compliance with Good Laboratory Practices (GLPs) and the product tested should be as comparable to the clinical product as possible (40 CFR 160 and 792, 1983). The following table (Table 4.4) lists typical studies conducted for the development of a typical biologic compared to a typical small-molecule drug.

There are specific preclinical challenges for developing biologics that differ from traditional small-molecule drugs. Some worthy of note here include: Species selection, study design, route, dose selection, and immunogenicity. Many of these topics were discussed in the original ICH S6 document of 1997 (International Council for Harmonisation, 1997c) or later in the 2011S6 Addendum (International Council for Harmonisation, 2011).

4.4.2.1 Species selection

One of the greatest challenges in the preclinical development of biotechnology-derived molecules is that of species specificity. Unlike traditional "small molecules," one cannot assume that a molecule will be active in the standard species used for toxicity testing. The lack of pharmacological activity in a species generally makes it unsuitable for use as in a toxicology study. Justification for using only one species can be based on the lack of a second species with biological activity or when the biological activity of the molecule or target toxicity can be adequately defined in just one species (ICH S6).

Table 4.4 Preclinical Toxicology Studies

List of Toxicology Studies for Biotech Product Development	List of Toxicology Studies for Small Molecule Development
Range finding studies	Screening studies
1-month studies	Range finding studies
3/6-month studies	Acute GLP studies
Safety pharmacology	1-month studies
Reproductive and development studies	Safety pharmacology
Tissue cross-reactivity studies (Mab)	Mutagenicity studies
Irritation/tolerance	3-month studies
Others as needed	Reproductive and development studies
	6-month rat
	9-month/1-year dog/monkey
	Industrial toxicology
	Diet RF studies
	Carcinogenicity studies

An example where the original species was not relevant is that of the early work performed on the interferons. In these cases, studies were performed in nonrelevant species and generated misleading information (Green & Terrell, 1992). Several studies were conducted in rodents with recombinant human interferons with little evidence of toxicity. This did not predict what was to occur in humans. Activity in monkey studies was more predictive, but there was the additional challenge of relative differences in the activity level between humans and nonhuman primates.

ICH S6 also refers to the use of homologous proteins and transgenic animals that express the human receptor. One example of a development program that relied on surrogates for safety assessment is that of infliximab (Treacy, 2000). Many of the challenges of these models are acknowledged in the ICH guideline. Animal models of disease can also be used with a strong scientific rationale.

4.4.2.2 Study design, route, and dose selection

Unlike most traditional pharmaceutical small molecules, biologics are not given once or twice a day, orally, in a pill or capsule. They are almost all dosed via a parenteral route and are not given (or "taken") daily. Some of these therapeutics are given in hospital settings or in clinicians' offices. Several are now self-administered approximately on a weekly basis. The dose regimen for the safety assessment studies should reflect the dosing regimen plan for the human studies. The dosing interval in the toxicology studies may be different depending on the half-life of the molecule in animals versus humans. Also, the development of antibodies in the animals may alter the pharmacokinetics, and modifying the dosing interval has been shown to reduce the incidence of antibodies.

Dose selection can be a challenge with biopharmaceuticals. Often the dose-limiting toxicity is related to the pharmacology (often referred to as exaggerated pharmacology) and it can be difficult to establish a margin of safety. The slope of the dose–response curve between the intended level of effect and an effect that leads to toxicity (even if it is the same effect such as an increase in hematocrit with erythropoietin-like molecules) may be very steep. The ability to produce formulations of proteins that allow for "high" doses can be difficult, especially if the protein formulation is viscous. Doses can be limited by practicality as well (e.g. dosing volume limits, maximal stable concentrations, etc.). There is always the desire to study a dose without effect, but this can be an additional challenge for molecules that are active at doses in the μg/kg level.

Other factors that need to be considered in dose selection are related to pharmacokinetics and drug metabolism. One specific area generally not deemed relevant for biologics is metabolism studies. Mass balance studies have not proved useful when performed with proteins. Radiolabeled studies have limited usefulness, as these labeled molecules undergo rapid metabolism and the label is often unstable. Protein biologics undergo proteolytic degradation into smaller peptides and then into individual amino acids. Distribution is viewed more from the perspective of target distribution, except for gene and cell therapies and viral or bacterial vectors.

4.4.2.3 Immunogenicity

It is generally well accepted that immunogenicity is not well predicted across species. It has also been well documented that many biological compounds intended for human use are immunogenic in animals (International Council for Harmonisation, 1997c). ICH S6 states: "Most biotechnology-derived pharmaceuticals intended for humans are immunogenic in animals." Traditional antigenicity studies or guinea pig anaphylaxis studies are not useful for predicting immunogenicity in humans and are now generally recognized as not being appropriate studies for biologics. When these studies were conducted years ago, at the request of some regulators, they were generally positive and lead to adverse effects in animals. Since there is little to no predictive value in these studies, and they were not considered appropriate, such studies have not been conducted since the publication of ICH S6.

All preclinical (and clinical) studies with biopharmaceuticals should include a characterization of immunogenicity and if there appears to be an impact on pharmacological activity, an assessment of whether these antibodies are neutralizing may be warranted. This may occur by either direct binding to an epitope with activity or by blocking the active site with steric hindrance by binding to a site in close proximity.

Assays to determine antibodies have become more sophisticated over the years and the newest technologies allow detection at lower levels than was achievable with traditional ELISAs. Clinically relevant antibodies include those that are clearing, sustaining, neutralizing, and/or cross-react with endogenous proteins. Antibody production alone, however, should not necessarily prohibit the conduct of these studies. The effect on pharmacokinetics and pharmacodynamics needs to be measured and evaluated.

4.4.3 CLINICAL DEVELOPMENT

In the last three decades, many biotechnology-derived products have been approved in the United States, Europe, Japan, and now globally. Innovations in molecular sciences have advanced research into a variety of therapeutics that target previously untapped ligands and receptors. These molecules and products hold the promise of treating unmet medical needs of patient with chronic and/or life-threatening diseases. The clinical development of these products is very similar to traditional small-molecule drugs and follows the same three study phases. Phase 1 focuses on safety, Phase 2 on efficacy and dose selection in patients, and then Phase 3 are generally well-controlled trials in larger populations. ICH Efficacy guidelines (E1–E16) are generally all applicable, including those focused on compliance and good clinical practices (GCPs) (International Council for Harmonisation, 2016).

There are a couple of unique aspects of biologics that can impact clinical development worth mentioning here. One is the difference in pharmacokinetics of these molecules. Many of these protein products, particularly monoclonal antibodies have very long half-lives, sometimes days, weeks, or even months. This can change the study design and monitoring plans for early studies. It can also impact the populations selected (normal healthy volunteers vs. patients) for Phase 1 trials. Some molecules can cause rapid adverse reactions such as infusion reaction and cytokine release phenomena, yet others can lead to adverse effects that take time to manifest, due to the longer pharmacological actions (Haller, Cosenza, & Sullivan, 2008). Due to the highly targeted nature of the molecules, most adverse findings are due to exaggerated pharmacology. The most common off-target toxicity is injection-site reactions.

Another unique concern with biologics is the formation of immunogenicity to the product itself. The more fully human a protein is, the less likely it is to be immunogenic, but there have been modifications in manufacturing and processing that have led to serious immunological consequences. Not all antibodies are clinically concerning. Some just bind without leading to serious effects. Others may lead to a more rapid clearing of the molecule, impacting the pharmacokinetics. The more serious antibodies are those that neutralize the effect of the product decreasing the intended efficacy of the biologic product. The most concerning are those antibodies that might cross reach with an endogenous protein as seen in clinical trials with megakaryocyte growth and development factor (MGDF) leading patients unable to generate platelets (Basser et al., 2002).

After the TeGenero incident (Horvath & Milton, 2009), the conduct of clinical trials with novel biologics has been modified to include slower dose escalation, sentinel dosing, and other safeguards to ensure such a disaster does not happen again. Safety reporting for biologics is also very important and is discussed more in the postmarketing section of this chapter.

4.5 THE REGULATORY PATHWAYS (HOW DO I BRING A NEW BIOLOGIC TO APPROVAL?)

A sponsor who wishes to begin clinical trials with a biologic must submit an investigational new drug (IND) application to the appropriate division of the FDA. The IND describes the biologic, its method of manufacture, and quality control tests for release. Also included is information about the product's safety in animal testing, as well as the proposed clinical protocol for studies in humans. For vaccines, the ability to elicit a protective immune response in animals is also included.

Premarketing (prelicensure) biologic or vaccine clinical trials are typically done in three phases, as is the case for any drug. Initial human studies, referred to as Phase 1, are safety and immunogenicity studies performed in a small number of closely monitored subjects. Phase 2 studies are dose-ranging studies and may enroll hundreds of subjects. Finally, Phase 3 trials typically enroll thousands of individuals and provide the critical documentation of effectiveness and important additional safety data required for licensing. At any stage of the clinical or animal studies, if data raise significant concerns about either safety or effectiveness, FDA may request additional information or studies, or may halt ongoing clinical studies.

4.5.1 INNOVATIVE AND EXPEDITED PATHWAYS

All of the expedited pathways in the previous chapter are applicable to biologics. These include Fast Track, Accelerated Approval, and Breakthrough Therapy Designation (see Chapter 3 for more details on expedited pathways). Breakthrough Therapy Designation is one of the more recent of the expedited pathways based on clinical data that provides drug development sponsors more frequent interactions with FDA. In addition, the FDA has recently developed a new designation: the Regenerative Medicine Advanced Therapy (RMAT) designation in 2017. Table 4.5 compares these designations. The RMAT designation was created as part of the 21st Century Cures Act. A drug is eligible for regenerative medicine advanced therapy (RMAT) designation if:

- the drug is a regenerative medicine therapy, which is defined as a cell therapy, therapeutic tissue engineering product, human cell, and tissue product, or any combination product using such therapies or products, except for those regulated solely under Section 361 of the Public Health Service Act and part 1271 of Title 21, Code of Federal Regulations,
- the drug is intended to treat, modify, reverse, or cure a serious or life-threatening disease or condition, and
- preliminary clinical evidence indicates that the drug has the potential to address unmet medical needs for such disease or condition.

Table 4.5 Comparison of RMAT and BT

	Breakthrough Therapy Designation (BT)	**Regenerative Medicine Advanced Therapy Designation (RMAT)**
Statute	Section 506(a) of the FD&C Act, as added by section 902 of the Food and Drugs Administration Safety and Innovation Act of 2012 (FDASIA)	Section 506(g) of the FD&C Act, as added by section 3033 of the 21st Century Cures Act
Qualifying criteria	A drug that is intended to treat a serious condition, AND preliminary clinical evidence indicates that the drug may demonstrate substantial improvement on a clinically significant endpoint(s) over available therapies	A drug is a regenerative medicine therapy, AND the drug is intended to treat, modify, reverse or cure a serious condition, AND preliminary clinical evidence indicates that the drug has the potential to address unmet medical needs for such disease or condition
Features	All fast-track designation features, including: actions to expedite development and review; rolling review Intensive guidance on efficient drug development, beginning as early as Phase 1 Organizational commitment involving senior FDA management	All breakthrough therapy designation features including early interactions to discuss any potential surrogate or intermediate endpoints Statute addresses potential ways to support accelerated approval and satisfy postapproval requirements

Most of the products fall into the Office of Tissues and Advanced Therapies (OTAT). The area of Regenerative Medicine is a rapidly expanding field that has the potential to treat serious conditions. CBER recognized the importance of these therapies and committed to help ensure they are licensed by developing the RMAT designation. A draft guidance was issued in late 2017. This guidance is intended to facilitate development and review of regenerative medicine therapies intended to address unmet medical needs in those with serious conditions. It includes cell therapies, therapeutic tissue engineering products, human cell and tissue products, and combination products using any such therapies or products, with a few exceptions. It may also include genetically modified cells that lead to a durable modification of cells or tissues that may meet the definition of a regenerative medicine therapy. Combination products may be eligible for RMAT designation if the regenerative medicine component provides the greatest contribution to the overall intended therapeutic effect.

Requests for both BT and RMAT can be submitted with the IND or after and ideally, no later than the end-of-phase 2 meeting. The FDA is required to respond within 60days after receipt of request. Both of these designations may be rescinded later in product development if the product no longer meets the designation-specific qualifying criteria (U.S. Food and Drug Administration, 2017e).

4.6 SUBMISSION AND REVIEW PROCESS

To place a medicinal product, new drug or biologic on the market in most countries, a company needs regulatory approval or marketing license. In the United States, this is done through the NDA or BLA process. In Europe, Marketing Authorization (MA) is either issued by the competent authority of country for its own territory (i.e. a national authorization) or can be granted for the entire Community (i.e. a community authorization).

4.6.1 UNITED STATES

As discussed in previous sections of this chapter, the US FDA regulates biologics. The regulating Center (CBER or CDER) will depend on the current classification of your biologic. The IND requirements (21 CFR Part 312) are all relevant, including format and content, financial disclosure, and clinical holds. The BLA process is similar to the NDA process discussed in Chapter 3, with the Establishment Licensing differences being noted in the previous sections of this chapter. In addition to the ICH guidelines already outlined, there are several FDA guidances and points to consider documents relevant to biologics that should be reviewed before embarking on the development of a new biologic. As many biologics are first-in-class, they may go to an FDA Advisory Committee for review and discussion during the BLA review process.

4.6.2 EUROPE

The overarching Regulatory Agency in the European Union (EU) is the European Medicines Agency (EMA). In the EU, the Centralized Procedure is *mandatory* for biotechnological medicinal products derived from recombinant DNA technology, controlled gene expression, or monoclonal antibodies. It is also mandatory for orphan drug products and advanced therapies (e.g. gene and cell therapies) making this the route by which most "biologics" will be regulated. Advanced Therapy Medicinal Products (ATMPs) have their own legislation found in Article 17 of Regulation (EC) No 1394/2007. In addition to the mandated use of the Centralized Procedure for these products, they must also go through the Committee for Advanced Therapies (CAT) for assessment.

A successful application under the Centralized Procedure delivers a single decision from the European Commission (EC) for marketing authorization valid throughout the European Union. The current framework of legislation for this procedure is Regulation No. 726/2004. This regulation lays out the procedure for the authorization of medicinal products. The objective of the Centralized Procedure is to provide major innovative drug products with direct access to a community-wide market. Products approved through the Centralized Procedure have a common Summary of Product Characteristics (SmPC or SPC; the EU equivalent of the US drug label) that must then be translated in all EU languages. Marketing authorizations are valid for 5years and then must be renewed.

One area of particular concern in Europe for biotechnology products was the association of animal-derived products with transmissible spongiform encephalopathy (TSE). Because of the seriousness of these diseases as seen in bovine spongiform encephalopathy (BSE) and Creutzfeldt–Jakob disease (CJD), EU legislation required applicants to demonstrate that a medicinal product is manufactured in accordance with strict Guidance (EMA/410/01 Rev3). The eventual goal is to replace ingredients of animal origin with vegetable-derived or synthetic materials.

The four principal parties involved in the Centralized Procedure review process include the company applying for the marketing authorization (MA), the EMA secretariat, the Committee for Medicinal Products for Human Use (CHMP), and the EC. The EMA appoints a project team leader for each application, who is responsible for all communication and correspondence with the applicant. The CHMP appoints a rapporteur and a corapporteur chosen from the CHMP members. The choice is based on the expertise required for the review of a certain therapeutic product area.

Release testing for products manufactured outside of the EU or European economic area (EEA) may also be required for every batch imported. It is therefore important to carefully consider the specifications selected for this testing. These tests can be more costly for biologics than for other products. For some biologics (e.g. vaccines), batch testing may be required even when produced within the EU or EEA.

4.6.3 REST OF WORLD

Most countries now follow ICH guidelines for the development of biologics. Because many biologics, unlike traditional small-molecule pharmaceuticals, are single sourced, developing countries generally rely on the review and approval of these products in the major jurisdictions (United States, EU, and Japan). They often require a certificate of pharmaceutical product (CPP) from the exporting country's Regulatory Agency (e.g. FDA) before importing the product into their country. They may also require lot release or batch testing locally on imported biologics. Some countries require that local clinical trials be conducted as well. It is important to investigate the local regulations specific for biologics in all countries where a company plans on marketing their product.

4.7 POSTMARKETING ACTIVITIES

There are several biologic-specific postmarketing activities worth mentioning. Many of these are related to the unique attributes of biologic manufacturing. They include lot releases, license revocation and suspension (related to Establishment licensure), recall orders (a provision of the PHSA for the recall of a specific batch lot or other quantity of a product, generally voluntary), and civil money penalties (rarely, if ever used). During the AIDS crisis, there were several court-ordered injunctions and consent decrees used against major US blood establishments.

In addition, the following activities are also worth noting:

- Postapproval manufacturing change supplements: For some changes in biologics manufacturing, clinical data may be necessary due to the complexity of the process.
- Establishment inspections (not specific to biologics) but because the manufacture is different, the FDA uses an approach called "Team Biologics" acknowledging the complexity of the manufacturing processes.
- New indications: New indications are filed as a supplemental BLA (sBLA). The regulations and processes for new indications and extensions are basically the same as for drugs (see Chapter 3).
- New delivery devices: Often as part of life-cycle management for therapeutic proteins or monoclonal antibodies, injection devices may be developed. For example, an injection pen to ease the self-injection process or reduce pain would be a combination product. The development of these types of products often involves multiple Centers at the FDA.
- Safety reporting for most biologics is similar to that for drugs, where adverse experience information must be reported on a timely basis depending on the seriousness and expectedness of the events. In addition, there are Biological Product Deviation Reports (BPDRs). This reporting also applies to contract manufacturers, distributors, and other vendors in the production and distribution

chain. This differs from drug manufacturing (field alert reporting). Of particular note is the Vaccine Adverse Event Reporting System (VAERS), which the FDA and the Centers for Disease Control and Prevention (CDC) co-manage. This system covers those vaccines that are part of the National Vaccine Injury Compensation Program (PHSA, tit. XXI, Subtitle 2). This system ensures that healthcare provides records and pertinent information for each vaccine administration.

4.8 SPECIAL TOPICS

Because biologics are not a homogenous category of products, whole chapters could be written about some of the special biologics (e.g. gene therapies or stem cells). Two specialty categories will be discussed in this section: Biosimilars and vaccines.

4.8.1 BIOSIMILARS

In the United States, the regulatory pathway for biosimilars was established through the Biologics Price Competition and Innovation Act (BPCI Act) as part of the 2009 Affordable Care Act. Biosimilars must undergo the same rigorous evaluation as the originator products to assure their efficacy, safety, and quality (U.S. Food and Drug Administration, 2017a). The first biosimilar in the United States was approved in 2015, Zarxio, a biosimilar of Neupogen (filgrastin). The development and approval of biosimilars in the United States started out slowly but has been picking up speed more recently. Biosimilars have been approved and marketed in Europe and other jurisdictions for several years now. Some of the challenges in the United States have been the longer patent lives, including the challenges of complicated manufacturing related patents. The next challenge will be meeting the hurdles of interchangeability, which has been set as a high standard by the FDA.

By definition, a biosimilar is highly similar to, and has no clinically meaningful difference in safety, purity, and efficacy from an existing FDA-approved reference product. This abbreviated licensing pathway, under Section 351(k) of the Public Health Service Act, permits reliance on certain existing scientific knowledge about the safety and effectiveness of the reference product and enables a biosimilar product to be licensed based on a reduced amount of preclinical and clinical data. The biosimilar must have the same mechanism of action, route(s) of administration, strengths, and dosage from(s) as the reference product. It can only be approved for indications and populations for which the reference product is already approved. The manufacturing facilities must meet the same FDA GMP standards. Comparative data are generated in order to evaluate the biosimilarity, not to independently establish the safety and efficacy of the new biosimilar product.

The pyramid below represents the way the FDA looks at the data on a new biosimilar. The FDA reviews the totality of the data and information, including the foundation of detailed analytical (structural and functional) characterization, animal studies, if necessary, then moving on to clinical pharmacology studies. If needed, other comparative clinical studies may be requested on a case-specific basis. This typically includes assessing immunogenicity, pharmacokinetics (PK), and, in some cases, pharmacodynamics (PD) and may also include a comparative clinical study.

The stronger the analytical data, the less preclinical animal work is likely to be required. For each product, the FDA determines what data are needed to demonstrate biosimilarity and which data elements can be waived if deemed scientifically appropriate. A biosimilar application can be modified based on the historical data on the reference product, and the data as it gets developed for the biosimilar itself. See pyramid (Fig. 4.1) from FDA web site on Biosimilar Applications (U.S. Food and Drug Administration, 2017a).

A biosimilar product may be approved for additional indications without direct studies of the biosimilar in each indication. If the totality of evidence in the biosimilar application supports biosimilarity in at least one of the reference product's indications, then it is possible for the manufacturer to use this data to "extrapolate" for additional indications. The biosimilar manufacturer and FDA need to scientifically justify approval for other indications that were not directly studied by the biosimilar manufacturer. The FDA evaluates all of the biosimilar product data to determine whether there are differences between the products that may impact the extrapolation to additional indications or populations. If no such differences are identified, approval of the biosimilar for other nonstudied indications or populations is generally supported. The FDA works with biosimilar manufacturers during product development, usually through a series of meetings, to determine what data might be needed to support extrapolation. It is generally unnecessary to require a biosimilar manufacturer to conduct clinical trials in all the same disease indications for which the reference product was originally studied and approved. This intent of this pathway is to provide more treatment options and potentially lowering healthcare costs. Since the data package required for approval of a biosimilar or interchangeable product is still extensive, the monetary savings is not comparable to small-molecule generics.

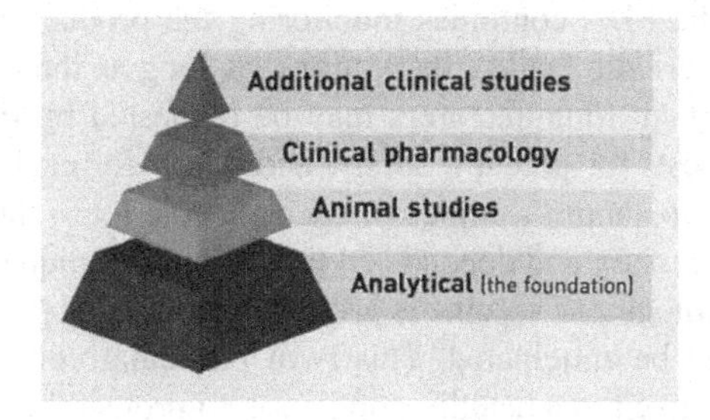

FIGURE 4.1

Biosimilar applications (https://www.fda.gov/ucm/groups/fdagov-public/documents/image/ucm581307.jpg). (https://www.fda.gov/Drugs/DevelopmentApprovalProcess/HowDrugsareDevelopedandApproved/ApprovalApplications/TherapeuticBiologicApplications/Biosimilars/ucm580429.htm#types).

4.8.2 VACCINES

Authority for the regulation of vaccines resides primarily in Section 351 of the Public Health Service Act and also in specific sections of the Federal Food, Drug, and Cosmetic Act. Vaccine clinical development follows the same general pathway as for drugs and other biologics (Stehlin, 1995). Clinical trials with a vaccine start after the submission and approval of an Investigational New Drug application (IND) to FDA. Similar to other biologics, the IND must describe the vaccine, its method of manufacture, and quality control tests for release. It also should include information about the vaccine's safety and ability to elicit a protective immune response (immunogenicity) in animal testing, as well as the proposed clinical protocol for studies in humans.

Premarketing (prelicensure) vaccine clinical trials are typically done in three phases, as is the case for any drug or biologic. As described previously, the studies usually follow the same three phases (Phase 1: Safety and immunogenicity; Phase 2: Dose–range finding studies; and Phase 3 often involves thousands of subjections to determine safety and efficacy over time) as with most drugs and biologics.

Like other biologics, successful completion of all three phases of clinical development can be followed by the submission of a biologics license application (BLA). Also during this stage, the proposed manufacturing facility undergoes a preapproval inspection during which production of the vaccine as it is in progress is examined in detail.

Following FDA's review of a license application for a new indication, the sponsor and the FDA may present their findings to FDA's Vaccines and Related Biological Products Advisory Committee (VRBPAC). This committee is made up of experts who provide advice to the Agency regarding the safety and efficacy of the vaccine for the proposed indication. Vaccine approval also requires adequate product labeling, to allow healthcare providers to understand the vaccine's proper use, including its potential benefits and risks, to communicate with patients and parents, and to safely deliver the vaccine to the public. With all of the publicity on vaccine safety in recent years, communication with the public (patients and parents) is key.

After licensure, the FDA continues monitoring the product and manufacturing activities, including periodic facility inspections, as long as the manufacturer holds a license for the product. Manufacturers may be requested by FDA to submit the results of their own tests for potency, safety, and purity for each vaccine lot. They may also be required to submit samples of each vaccine lot to the FDA for testing. This is not always necessary and depends on the testing techniques and history.

Until any drug, biologic, or vaccine is given to a broader population, all potential adverse events cannot be anticipated. This is of particular concern with vaccines, as they are generally given to a healthy population to protect them from developing serious diseases. Thus, many vaccines undergo Phase 4 safety monitoring studies once marketed. Also, the government relies on the Vaccine Adverse Event Reporting System (VAERS) previously described to identify problems after marketing begins (Parkman & Hardegree, 1999).

4.9 CONCLUSIONS

Biologics and the biotechnology (now the biopharmaceutical) industry have grown tremendously. In the 1980s, there were very few INDs for these products per year. By 1990, there were over 200 open INDs and now there are over 350 biologics in late stage development. There are over 250 biotechnology healthcare products and vaccines available worldwide (Biotechnology Innovation and Organization, 2017). Sales of biologics were estimated to be about $93 billion globally in 2009 (McCamish & Woollett, 2011). Biotechnology is now mainstream.

REFERENCES

42 United States Code 262(i). (n.d.). *Regulation of biological products.*

Basser, R. L., O'Flaherty, E., Green, M., Edmonds, M., Nichols, J., Menchaca, D., et al. (2002). Development of pancytopenia with neutralizing antibodies to thrombopoietin after multicycle chemotherapy supported by megakaryocyte growth and development factor. *Blood, 99*(7), 2599–2602.

Biotechnology Innovation and Organization. (2017). *What is biotechnology*. Retrieved from https://www.bio.org/what-biotechnology.

Cosenza, M. E. (2010). Safety assessment of biotechnology-derived therapeutics. In S. C. Gad (Ed.), *Pharaceutical sciences encyclopedia: Drug discovery development, and manufacturing* (pp. 1–17). Hoboken, NJ: Wiley.

Green, J. D., & Terrell, T. G. (1992). Utilization of homologous proteins to evaluate the safety of recominatnt human proteins—Case study: Recombinant human interferon gamma (fhIFN-y). *Toxicology Letters, 64/65*, 321.

Haller, C. A., Cosenza, M. E., & Sullivan, J. T. (2008). Safety issues specific to clinical development of protein therapeutics. *Clinical Pharmacology & Therapeutics, 84*(5), 624–627.

Horvath, C., & Milton, M. (2009). The TeGenero incident and the Duff report conclusions: A series of unfortunate events or an avoidable event? *Toxicologic Pathology, 37*, 372–383.

International Council for Harmonisation. (1997a). *Q5A: Viral safety evaluation of biotechnology products derived from cell lines of human or animal origins*. International Council for Harmonisation.

International Council for Harmonisation. (1997b). *Q5D: Derivation and characterization of cell substrates used for production of biotechnological/biological products*. International Council for Harmonisation.

International Council for Harmonisation. (1997c). *S6: Preclinical safety evaluation of biotechnology derived pharmaceuticals*. International Council for Harmonisation.

International Council for Harmonisation. (2011). *Preclinical safety evaluation of biotechnology-derived pharmaceuticals S6 (R1)*. International Council on Harmonization.

International Council for Harmonisation. (2016). *Integrated addendum to ICH E6 (R1): Guideline for good clinical practice E6(R2)*. International Council on Harmonisation.

Korwek, E. L., & Druckman, M. N. (2015). Human biologics. In D. C. Adams, R. M. Cooper, M. J. Hahn, & J. S. Kahan (Eds.), *Food and drug law and regulation*, (3rd ed., pp. 513–551). Washington, DC: FDLI.

McCamish, M., & Woollett, G. (2011). Worldwide experience with biosimilar development. *mAbs, 3*(2), 209–217.

Parkman, P. D., & Hardegree, M. C. (1999). Regulation and testing of vaccines. In S. A. Plotkin, & W. A. Orenstein (Eds.), *Vaccines* (3rd ed., pp. 1131–1143).

Stehlin, I. B. (1995). How FDA works to ensure vaccine safety. *FDA Consumer Magazine*.

The European Parliament and The Council of the European Union. (2007). *Eur-Lex. regulation (EC) No 1394/2007 of the European Parliament and of the Council of 13 November 2007 on advanced therapy medicinal products and amending Directive 2001/83/EC and Regulation (EC) No 726/2004*. Retrieved from http://eur-lex.europa.eu/legal-content/EN/ALL/?uri=CELEX%3A32007R1394.

Treacy, G. (2000). Using an analogous monoclonal antibody to evaluate the reproductive and chronic toxicity potential for ahumaized anti-TNFa monoclonal antiobody. *Human & Experimental Toxicology*, *19*, 226.

U.S. Food and Drug Administration. (2009). *Public Health Service Act*. Retrieved from https://www.fda.gov/regulatoryinformation/lawsenforcedbyfda/ucm148717.htm.

U.S. Food and Drug Administration. (2017a). *Biosimilars*. Retrieved from https://www.fda.gov/drugs/developmentapprovalprocess/howdrugsaredevelopedandapproved/approvalapplications/therapeuticbiologicapplications/biosimilars/default.htm.

U.S. Food and Drug Administration. (2017b). *CFR-Code of Federal Regulations Title 21*. Retrieved from https://www.accessdata.fda.gov/scripts/cdrh/cfdocs/cfcfr/cfrsearch.cfm.

U.S. Food and Drug Administration. (2017c). *Healthcare providers (biologics)*. Retrieved from https://www.fda.gov/BiologicsBloodVaccines/ResourcesforYou/HealthcareProviders/default.htm.

U.S. Food and Drug Administration. (2017d). *Part 58 good laboratory practice for nonclinical laboratory studies*. Retrieved from https://www.accessdata.fda.gov/scripts/cdrh/cfdocs/cfcfr/CFRSearch.cfm?CFRPart=58.

U.S. Food and Drug Administration. (2017e). *Regenerative medicine advanced therapy designation*. Retrieved from https://www.fda.gov/biologicsbloodvaccines/cellulargenetherapyproducts/ucm537670.htm.

FURTHER READINGS

International Council for Harmonisation (1995). *Q5B analysis of the expression construct in cells used for production of r-DNA derived protein products*. International Council for Harmonisation.

International Council for Harmonisation. (2004). *Q5E: Comparability of biotechnological/biological products subject to changes in their manufacturing process*. International Council for Harmonisation.

International Council for Harmonisation. (1995). *Q5C: Stability testing of biotechnological/biological products*. International Council for Harmonisation.

U.S. Food and Drug Administration. (2013). *The road to the biologic revolution—highlights of 100 years of biologics regulation*. Retrieved from https://www.fda.gov/AboutFDA/WhatWeDo/History/CentennialofFDA/CentennialEditionofFDAConsumer/ucm096141.htm.

Yang, L. (2013). Biologics submission. In P.A. Jones (Ed.), *Fundamentals of US regulatory affairs* (pp. 273–298). Rockville, MD: Regulatory Affairs Professionals Society.

CHAPTER

5 Medical device and diagnostic products

Frances J. Richmond*, Sai S. Tatavarty**

**International Center for Regulatory Science, University of Southern California, Los Angeles, CA, United States; **Abbott Diabetes Care, Alameda, CA, United States*

"At some point in every person's life, you will need an assisted medical device - whether it's your glasses, your contacts, or as you age and you have a hip replacement or a knee replacement or a pacemaker. The prosthetic generation is all around us."

– Aimee Mullins

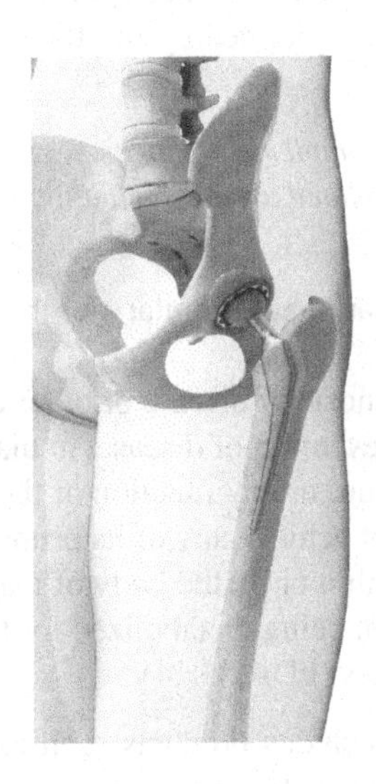

An Overview of FDA Regulated Products. http://dx.doi.org/10.1016/B978-0-12-811155-0.00005-3

CHAPTER OBJECTIVES

After reading this chapter, the reader will be able to:

- identify what constitutes a medical device,
- identify how devices are classified in the United States and certain other major markets,
- describe the corresponding regulatory paths to commercialization for different product classes,
- identify key activities that must be carried out pre- and postmarket,
- describe ways in which the FDA is attempting to facilitate innovation, and
- differentiate approaches to in vitro diagnostics as compared to medical devices.

5.1 REGULATING DEVICES IN THE UNITED STATES

5.1.1 DO I HAVE A MEDICAL DEVICE?

Devices enter the regulated environment when they cross the boundary from being a consumer product to a product intended to manage a medical condition. In the United States, a medical device has been defined by the Food, Drug, and Cosmetic Act of 1938 as...

> *... an instrument, apparatus, implement, machine, contrivance, implant, in vitro reagent, or other similar or related article, including a component part, or accessory which is:*

- recognized in the official National Formulary, or the US Pharmacopoeia, or any supplement to them,
- intended for use in the diagnosis of disease or other conditions, or in the cure, mitigation, treatment, or prevention of disease, in man or other animals, or
- intended to affect the structure or any function of the body of man or other animals, and which does not achieve any of its primary intended purposes through chemical action within or on the body of man or other animals and which is not dependent upon being metabolized for the achievement of any of its primary intended purposes (FDA, 2014).

This definition obviously includes products such as pacemakers and catheters. Less obvious are items like diagnostic test kits, such as pregnancy test kits, and i-phone applications such as those that record blood sugar levels to guide insulin dosing in diabetics.

The "medical device" definition is important because it not only shows where to place the line between medical and nonmedical products, but also the line between devices and other medical products such as drugs and biologics. A key phase, "which does not achieve any of its primary intended purposes through chemical action," identifies that most devices will replace a structural or mechanical function. However, some medical devices replace or alter electrical functions, such as heart rhythms or brain function. Another key phrase, *including a component part, or accessory,* identifies that items such as liquid reagents will be considered as devices even though at first glance this might seem surprising. Thus, a diversity of products, from contact lens cleaning solutions to oral medications, might also be classified as devices either because they are accessories to a device or because they do not act by entering the body through the gut wall or the skin. In vitro diagnostics, tests that are conducted outside of the body to identify components of the blood or urine, for example, are also covered by the definition of a medical device and are discussed later. A treadmill would be a medical device, too, if it claimed a use to treat a disease, such as treatment of patients with coronary infarction. What the definition excludes are devices used to improve wellness. A treadmill used for exercise in healthy individuals would not be a medical device. Whether a device is a medical device depends on what it claims to do. Thus, before developing a strategy for a device, it is important to make sure that the product under study indeed fits into the Medical Device category.

Medical devices can also be part of what has come to be known as "combination" products. Combination products have components that fit into two different regulatory divisions, by combining the device with another type of regulated product, such as a drug or biologic. When the principal mode of action of the combined product is device-like, then the regulatory filings of the combination product will be managed by the Center for Devices and Radiological Health (CDRH) that will then be called the "lead" Center. The regulations for devices will be followed, but additional requirements related to the secondary component will also need to be met to support the regulatory submission. Conversely, when the primary action is to achieve a pharmaceutical purpose, then the regulatory oversight of the combination product will be managed through the Center for Drug Evaluation and Research (CDER), or through the Center for Biologics Evaluation and Research (CBER), but regulatory requirements appropriate for the device components will also need to be met.

5.1.2 HOW DO I CLASSIFY MY DEVICE?

Medical devices vary from simple products such as bedpans to complex electronic devices implanted in the brain. The amount of oversight and testing required to assure the safety and efficacy of a device should be tailored to its complexity and function. This challenge was recognized when the Medical Devices Amendments of 1976 and associated regulations established a risk-based classification system to govern medical devices. In this system, devices with lowest risk were categorized as Class I devices with relatively straightforward requirements whereas those with the highest risk were categorized in Class III with more stringent requirements.

The requirements under different Class designations take into account the fact that all devices, regardless of class, must be safe and effective, and that this can only be assured by assessing risks, designing products that mitigate those risks and manufacturing products according to well-defined specifications. All devices, regardless of Class, are subject to what has come to be called "general controls." These include establishment and listing requirements. Under these requirements, all device companies marketing in the United States must register their establishments with the Food and Drug Administration (FDA); if the company is located outside of the United States, it must have an authorized representative in the United States to serve as the legal representative with whom the FDA can interface. The device companies also must list their products on the FDA web site. The fees for registration and listing are typically a few thousand dollars and change yearly. Medical device manufacturers must also label their products in a truthful and appropriate way, manufacture in compliance with certain parts of Quality Systems Regulations (QSRs) (see Chapter 12), maintain records and complaint files, and report adverse events. To this list, other requirements, called "special controls", may be added for certain types of products, particularly for products in classes higher than Class I.

5.1.2.1 Class I devices

Class I devices are low-risk products that have the fewest regulatory requirements. They must adhere to "general controls," as described above. A small number of devices in Class I, such as patient restraints, have proved historically to be more risky than initially anticipated when they were put into Class I. These devices are additionally required to submit a premarket notification submission to the FDA, like that usually demanded for Class II products as described below. A small number of Class I devices require "design controls," also described below. A listing of Class I devices with additional requirements can be found on the FDA web site. For some Class I devices, however, many of the usual FDA submissions and quality systems requirements are waived; these waivers are also identified on the FDA web site. Thus many, but not all, Class I products can go directly to market with no more than establishment registration and product listing and a modest set of quality and labeling requirements.

5.1.2.2 Class II devices

Class II devices are considered to be devices with greater risks, but the risks can be mitigated to an acceptable level by assuring that the product meets certain performance standards or requirements, called "special controls." To assure that such controls have been satisfied, most Class II devices can only be marketed after providing FDA with documentation in the form of a "premarket notification," also called a "510(k)" submission. The 510(k) term refers to the location of the rule governing the premarket notification, in Section 510(k) of the Food, Drug, and Cosmetic Act. The content requirements for the 510(k) premarket notification are described in 21 CFR 807.87. Recently, the FDA has begun to review whether all Class II devices should be required to use a prenotification route and have exempted some low-risk products from the premarket notification requirements.

A premarket notification, as its name implies, is a relatively short submission that is used by a company to *notify* the FDA about its intent to market a product. The premarket notification must be submitted to the FDA at least 90 days before its intended marketing date so that the FDA has time to review the submission. The FDA can examine the information provided by the company and then delay or prevent the product from being commercialized if concerns exist. It is common for FDA to ask questions that must be satisfied before the product can go to market. The regulatory "time-clock" stops while responses from the company are prepared. Companies should therefore anticipate a review time that may run longer than 90 days. However, FDA does not review premarket notifications with the same level of intensity as it does submissions for Class III products. The process is therefore not considered to be an approval but rather a "clearance" process.

The clearance of a Class II product depends on the ability of that device to satisfy performance standards that are specified for that particular product group. For example, a measuring device would have requirements related to accuracy and precision, whereas a sterile device would have requirements related to an absence of viable microbes. In many cases, the performance standards can be satisfied by engineering or animal tests. For example, an implantable Class II product such as an intervertebral disc replacement might need to undergo biocompatibility testing. It is important to understand the types of testing that a product must undergo. For example, most products for use in the body have requirements to assure biocompatibility, according to the standard ISO 10093-1. However, such requirements can change with time. For example, certain types of biocompatibility testing that once were required on all implanted products may now be exempted if the materials used in the device have such a long history of safe use that the FDA sees little reason to repeat the tests on that type of material. Often insights into the necessary testing can come by examining the testing done on other such products, searching for specific guidance documents or consulting standards relevant to that product type or certain types of components, such as power supplies or software.

Studies in humans are required to support claims of performance or safety for only a minority of Class II products. For the rest, requirements can be satisfied by bench and animal testing. Class II products also must be produced under a specific set of quality regulations called quality systems. As part of that approach, the product is subject to "design controls," an approach to managing design in a controlled and documented way that is explained in more detail below. Because every device is a little different, the special requirements for a particular product must be understood during the design phase, so that the correct set of tests can be completed efficiently as the product design matures.

The 510(k) submission process is further specialized because it requires that the company identifies an already marketed predicate product to which the new product must show "substantial equivalence." The need to prove substantial equivalence to a predicate has historical roots that make US submissions different from those in almost every other country in the world. It was put into the law because of political pressures when the original FDA rules for devices were developed, in the Medical Device Amendments to the FD&C Act, in 1976. At that time, attempts to regulate medical devices were opposed by device manufacturers who were unhappy at the prospect that they would have to prepare formal submissions to the FDA for products that were already on the market and had performed safely, often for many years. To satisfy those concerns, "preamendment" devices in Class II were "grandfathered", that is, allowed to stay on the market, without a regulatory submission. In response, manufacturers with devices that were not yet on the market but in the pipeline were unhappy that they might have to satisfy a new set of rules that were not being applied to their competitors already on the market. To create a more equal playing field, the new manufacturers were allowed to avoid a more lengthy premarket approval process by submitting evidence that their new product was "substantially equivalent" to a product that had already been on the market. By allowing a company to demonstrate substantial equivalence to a predicate, the FDA acknowledges the company's claim that its new product is as safe and as efficacious as its predicate. However, that "clearance" does not guarantee that either the new product or its predicate is necessarily safe or efficacious. The subtlety of clearance versus approval is often not recognized by some inexperienced companies with Class II products, so it is not surprising that some manufacturers claim incorrectly that their 510(k) submissions have been "approved" by the FDA.

An interesting problem arises with this particular regulatory path when a cleared and marketed Class II product undergoes a change. There is no mechanism to update a 510(k) application by supplying supplemental data. Thus, a changed product may need to go through a new premarket notification process. Confusion can be created when trying to decide whether a certain type of change will trigger the need for a new 510(k) submission and FDA has provided guidance documents on this issue. Typically, a new 510(k) is needed if the product is modified in a way that could affect its safety and effectiveness. However, it can be tedious for companies and burdensome to the FDA if traditional 510(k) submissions are required every time a product is modified. Thus, a couple of shortened pathways exist for situations in which changes can be handled in a simplified way. The first is the situation in which relatively simple modifications do not change the intended use or basic technological features of the device. In such a case, the company making the device that is already on the market can supply a summary of the change by providing a "Special" 510(k) submission. This focused submission includes the results of a risk analysis related to the change and then describes the way in which the design control system was used to assure that the device continues to meet its originally stated requirements related to safety and efficacy. The Special 510(k) process thus is used only by the company that makes the original device and is a way to allow for continued modifications in response to market demands or changing technological opportunities and challenges. Another, less often used, pathway to shorten the 510(k) process is through the development of an "abbreviated" 510(k) submission. In an "abbreviated" submission, the company can submit a summary report rather than a detailed evaluation of testing if it adheres to relevant device-specific guidance documents, special controls, or relevant consensus standards. Both of these routes not only simplify the regulatory documentation required for submission but also help to reduce the review obligations of the FDA, potentially shortening the review process. Whatever the route to approval, the submissions need to be managed so that the progressive iterations are organized in a systematic archive.

5.1.2.3 Class III devices

The riskiest devices are subject to a much higher level of scrutiny by the FDA. Class III devices make up only about 10% of the device population, but their review requires a disproportionate amount of testing and analysis. Before a Class III device is marketed, it must be preapproved by the FDA. This preapproval is sought by submitting a premarket approval application, or PMA, to the FDA, many months in advance of the planned commercialization launch. The rules governing the PMA are described in 21 CFR 814.20. The FDA typically expects to take about 10 months to review a PMA application, although a device intended for a serious or life-threatening disorder without alternative options for treatment would have access to a priority pathway that would be reviewed in about 6 months. Their regulatory examination is designed to "provide a reasonable assurance" that the device is safe and effective for the use that is claimed on its label considering the risks that it may pose to patients.

A premarket approval application contains a comprehensive catalog of the design, testing, manufacture, and labeling associated with a device that could pose substantial risks to patients. Thus, it must contain the materials needed to support the arguments that the design is well-considered, the testing is thorough, and the manufacturing is of high quality and under stringent control. When the submission is reviewed, the regulatory agency will be trying to understand if the risks posed by the device are controlled appropriately. The submission will generally include the methods and results of clinical trials and will provide draft labeling to include all of the elements required by the Agency, including, for example, its name, indication for use, unique identification code, shelf-life, sterility status, and company address (Fig. 5.1). Most Class III devices will also provide the physician or other users with an "Instructions for Use" document, to be evaluated as part of the review.

A common statement when answering questions about what goes into a regulatory submission is, "it depends." For example, the large majority of PMA applications for an implantable device will contain clinical trial data. However, exceptions can occur. For example, Advanced Bionics gained approval for a spinal cord stimulator in 2004 by describing the clinical studies carried out by others, with the argument that the new stimulator was very similar to others on the market. Presumably, nothing would be gained by conducting clinical studies if the device could be shown to meet the same performance criteria as other similar devices, an argument that is already made commonly for 510(k) products.

Unlike devices approved by premarket notification, products that are approved through the PMA route do not have to provide a new PMA each time the company makes a change to the device design or labeling. Instead, changes to the approved

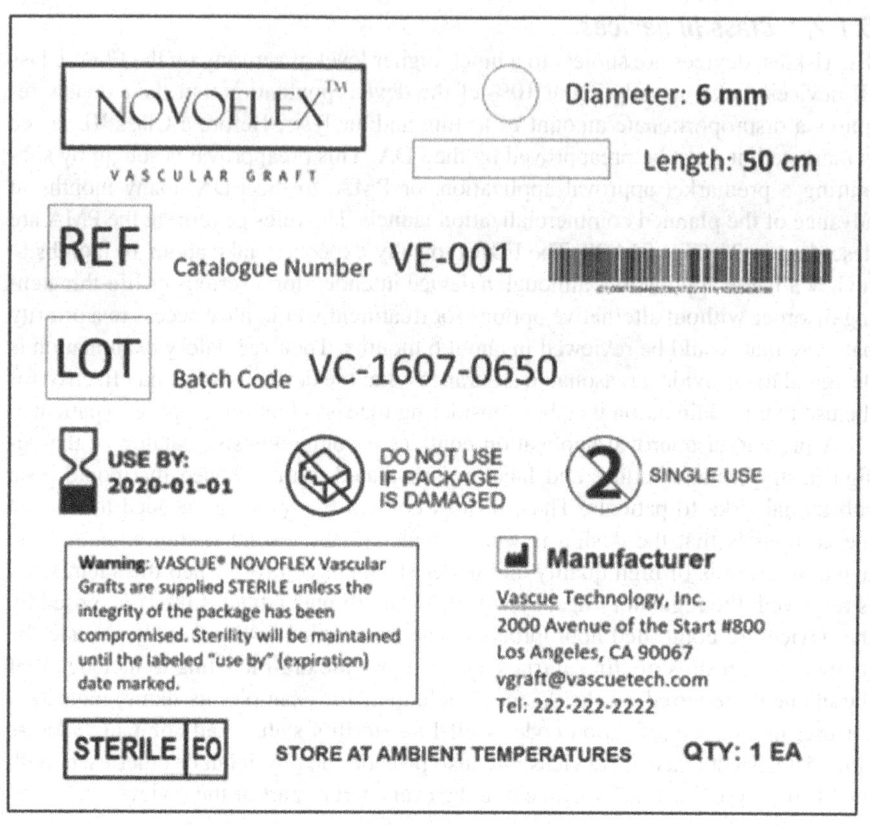

FIGURE 5.1 Example of draft labeling for a 510(k) submission.

Source: MSc student project, USC.

device are made by submitting a "supplement" to the PMA dossier that describes the relevant change. A supplement is also used when new information is to be added to the approved PMA. For example, results of a clinical postmarketing study might be added to the device file in this way. However, if information is added to a PMA or a PMA supplement *during* the course of its review and *before* it is approved, the new information is provided as an "amendment."

Managing the regulatory documentation associated with a Class III product over its lifetime is a challenging task. The submissions of applications, supplements, and amendments must be organized in controlled document files that maintain the revision history. Further, devices that are PMA approved also have a number of other reporting obligations. The PMA sponsor must provide annual reports at yearly intervals from the date that the PMA was approved, to update the FDA with a summary of the device experience and changes over the last year. It enumerates the adverse events and product problems that have been seen over the year and explains more modest changes that have been made to the manufacturing, design, or labeling that do not require a supplement to the PMA. Some of these changes would have been reported separately under the rules for reporting serious adverse events and malfunctions. However, the annual report may also contain less serious changes that do not affect the safety or efficacy of the device. The annual report also communicates some types of new information such as information gleaned by surveying the literature regarding the use of the device. Each annual report is also considered to "supplement" a PMA. A device that has been on the market for many years may have hundreds of supplements to be tracked in the company's document control system.

5.1.3 HOW DO I FIND OUT MY DEVICE CLASS?

It can be easy or difficult to find out the device class for a particular product. Devices that are already on the market are placed into groups and each group is governed by a specific regulation. The challenge is to find the regulation that matches the new device under consideration. A database provided by the FDA, called the "product classification" database, is useful but can be frustrating if the FDA uses a different word to describe the device than the company might use. For example, searching for an arthroscope is easy. The FDA designates an arthroscope as an arthroscope and so does the company, so entering the word "arthroscope" into the "device name" field gives you immediately the regulation that is needed. However, not all devices are easy to find this way. Let us look at the example of a false eye. The search term, "false eye" or "eye prosthesis," yields no results. The search term, eyeball, gives one result but it is for a "Conformer, Ophthalmic, Tissue," defined as a "molded plastic and biological tissue insert that is introduced temporarily between the eyeball and eyelid to maintain space in the orbital cavity." However, the term, "artificial eye," yields the desired regulation for "Eye, Artificial, Non-Custom."

Sometimes it is easier to search for the appropriate classification of your device by finding the regulation matching the device of another company with a similar product. You can do this by searching on the specific company and product name or by using a generic word that would bring up several device categories, such as "eye." It is important to find the specific regulation appropriate for your particular product, because different products may have deviations from the general rules described above that can affect the regulatory path of the device. For example, a few Class III devices do not require a premarket approval submission and can be submitted through a 510(k) route; a few Class II devices are exempt from premarket notification while a few Class I products require premarket notification. Also helpful is the fact that certain groups of products under a single regulation are also given a three-letter subclassification that further narrows the device groups in that subclass. For example, the regulation governing an "orthopedic manual surgical instrument," 21 CFR 888.4540, contains over 40 subclasses, such as "staple driver," subclass HXJ, or "wrench," subclass HXC.

Occasionally, it seems impossible to find a description that fits your device in the regulations. This problem may happen for two reasons. It may be that the device is really there but the search has been insufficient or the device has an unclear classification. The FDA has a process, for a fee, to request a classification determination. Such a determination is typically made within 60 days and requires a letter describing the device. However, your device may be so new that it does not yet have a predicate or governing regulation. This kind of product is more challenging to manage. Without a predicate, it will be designated automatically in Class III, where it would require a premarket approval application prior to commercialization. Such a Class designation may not be appropriate for a less risky device, so there is a route by which the device can be "admitted" to a more appropriate class, by what is called a "de novo" classification procedure.

5.1.4 WHAT IS A DE NOVO SUBMISSION?

A de novo submission aims to establish the classification of an unclassified device for future regulatory purposes into a class appropriate for the risk profile of that type of device. When a device is down-classified to Class II, the FDA is confirming that the risks of the product can be managed through special controls and the device can then serve as a predicate of other such products as they enter the market. If it is down-classified to Class I, it is confirming that it can be handled through the application of general controls. The reclassification process involves a petition to the FDA that outlines the rationale for the down-classification, and, for class II devices, a submission that resembles a 510(k), to which the FDA should respond within 120 days. After that reclassification, other similar devices can then be submitted for approval by the premarket notification process if Class II or can institute general controls and be placed on market without a further submission if it is put into a Class I category for which the premarket notification has been waived.

5.1.5 REDUCING THE BURDEN FOR PRODUCTS WITH A SMALL MARKET

Most medical devices will be sold to thousands of patients. These sales are important because they allow the company to recoup its investment in development. However, some medical products affect only a small number of patients, so that devices for this population often cannot be developed in a way that assures sufficient financial return to justify the costs of development and manufacture. Not surprisingly, then, manufacturers are reluctant to develop devices for rare diseases. To provide incentives that might help to reduce this hurdle, Congress passed the Safe Medical Devices Act of 1990. This Act authorized the FDA to develop regulations for "humanitarian use devices" (HUDs). FDA was given the authority to issue a final rule defining humanitarian devices as medical devices to treat or diagnose a disease or condition "that affects or is manifested in fewer than 8000 individuals in the United States per year." FDA defined a two-step process to facilitate the regulatory approval of such devices. In the first step, the device is designated as an HUD by the Office of Orphan Product Development (OOPD). The request for designation is relatively simple. It identifies the proposed indication and the reason why the device is needed. It then must make the argument, using authoritative references, that the device will serve a population of fewer than 8000 individuals. A company can sometimes obtain a designation if the patient group is larger but the device is only appropriate for a subset of that patient group called a "medically plausible subset." For example, it might be the case that a disease state such as spinal cord injury would affect many more than 8000 patients but only a small subgroup would be untreatable by current drug and device approaches and would be effectively treated by a new device. In such a case, the small subset would be the only subgroup in that disease state whose members would be appropriate candidates and would form a "medically plausible" subset.

In a second step, the company completes the required design and testing so that it can submit an application, called a Humanitarian Device Exemption (HDE), to gain marketing approval from CDRH. This marketing application is like a shorter version of a PMA application. The content requirements for HDE are described in 21 CFR 814.104. It must provide a reasonable amount of evidence to argue that the product does not pose a significant risk of illness or injury, but it does not have to prove effectiveness. Instead, the company must provide evidence that the "probable benefit" to health outweighs the risk of using the device. Most commonly, this evidence is provided by some combination of the literature review, bench testing, in vitro and animal testing, and human clinical study. For each unique device, it is important to work with the FDA to decide on the types of evidence that will be acceptable as a basis for the product's approval. The application must show that no comparable devices are marketed to treat or diagnose the same medical problem other than another device approved under HDE pathway. Once a comparable device for the same indication is marketed through either the PMA or the 510(k) process, the HUD will no longer be justified as an approval route and the HUD will be withdrawn.

In the past, limits were placed on the amount of money that companies with HDE approvals could charge. If priced higher than $250, they had to justify their price to show that the sales of the product did not result in income higher than that required to recover the costs of research, manufacturing, packaging, and distribution. The company had to attest that the price would not exceed the costs of making the product, as stated above, by providing a report to that effect from an independent certified public accountant. However, this rule was revised under the Pediatric Medical Device Safety and Improvement Act of 2007 to exempt pediatric devices and certain devices that treat both pediatric and nonpediatric patients from price justifications. Such eligible devices are given an annual distribution number (ADN) that limits the number of sales of the device that can be sold, but does not limit price. HDEs, like premarket approvals, come with the requirement to provide an annual report after the approval is given. This report will include information on the number of devices that have been distributed. Should the number of devices that are sold beyond the ADN, the sales must be reported to FDA. Products sold in excess of the ADN are subject to the prohibition on profits that is placed on other types of HUDs, unless the FDA agrees to increase the ADN. FDA could also revoke the HUD designation if the company later proved to violate the applicable rules.

HUDs are often marketed after very little clinical study. This situation exposes patients to some level of additional risk, so FDA has placed certain additional restrictions on the use of HUDs. First, the use of the HUD must be approved by the Institutional Review Board (IRB) of the facility in which the devices are to be used. Because HUDs are *approved* rather than *investigational* products, patients do not need to give informed consent to be treated with the device. Some IRBs, however, may decide that they require some form of patient approval or acknowledgement before an HUD is implanted or used. Second, the HUD must be labeled explicitly as an HUD whose effectiveness for the approved indication has not yet demonstrated.

5.2 RECENT INITIATIVES TO ENCOURAGE INNOVATION

The regulatory agencies are acutely aware of the fact that long regulatory review periods can slow access to important new therapies and particularly therapies that can offer help to serious or life-threatening diseases for which no good treatment exists. For several years, FDA has given priority to new products that address an unmet need for a life-threatening or seriously debilitating condition. Recently, a new pathway, called the Innovation Pathway, has also been introduced to accelerate the testing and review of such novel devices. This new program will make use of a new group called the Center Science Council that will oversee the use of the Innovation Pathway to assure that any hurdles to efficient review and approval are managed effectively. FDA will also try to identify experts for that product and will assign it to a case manager who will assist the company to build a documented roadmap with flexible clinical trial protocols and established review timeframes. However, regardless of the efforts that regulators try to make to shorten development times, innovative products can have specialized design features and clinical challenges that complicate their development. Many questions may need to be answered before the regulators can be convinced of safety and efficacy of a truly innovative product, especially one with significant risk.

5.3 REGULATORY STRATEGY

Companies that are investing much time and money into a new product must develop a regulatory strategy in much the same way as it develops a business strategy. Indeed, the regulatory requirements can dictate the business strategy for the product. It is therefore important to view a product not just as an invention but also as a device with a life cycle that has a number of phases, as shown in Fig. 5.2.

5.3.1 MANAGING THE DESIGN PHASE

Until the 1990s, regulators and companies paid relatively little attention to "controlling" the design phase. For most low-risk devices in class I, this lack of attention

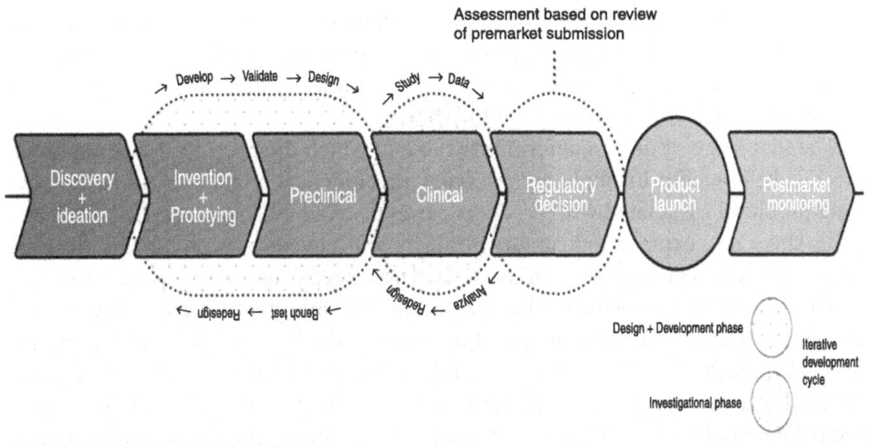

FIGURE 5.2 Medical device lifecycle.

Source: FDA, Medical Innovation Initiative White Paper, 2011.

was understandable because the potential hazards associated with the product were typically obvious to the manufacturer and user. However, for higher risk devices, this lack of attention resulted in safety problems that have killed or injured many people. A commonly used example that led to increased regulation of devices is a heart valve made by Shiley in the early 1980s that was reengineered in a way that weakened the struts holding the valve. The valve subsequently was plagued by serious adverse events when some of the struts fatigued and broke after persistent use. Efforts were therefore made to control the risks of design by revising the good manufacturing practices (GMPs) that had been required for such products in a way that extended the oversight for devices to include design under a system of activities called "design controls." These revisions converted the GMPs into a more comprehensive set of requirements called "Quality System Regulations" (QSRs, 21 CFR 820). Under QSRs, design controls became required for all Class II and III devices and for a small number of Class I devices considered to have design-related safety risks.

Design controls are a set of steps used to assure that device design and testing are planned, executed, and documented systematically. Activities associated with each step are brought together in a "Design History File" (DHF), a set of reference documents that demonstrate how the design has been developed and controlled. Design activities are organized in the DHF according to procedures that must be put into place by the company. The DHF typically grows as design activities advance and certainly must be in place before a clinical trial if such a trial is to be conducted.

When it becomes clear that an invention is going to be developed as a commercialized device, a plan is created to organize its design activities into a series of steps. A first step of design control is "design and development planning." This step involves the explicit identification of the job functions and deliverables for different phases. It identifies when and how reviews will be scheduled and serves as a project management guide for the subsequent design activities. If the product is a new Class III product, this document will eventually be fed into the PMA submission to the FDA. Most commonly, design controls are organized according to a "waterfall" framework illustrated in Fig. 5.3.

The first stage of design control is a stage in which the user needs are captured. This can be done, for example, by interviewing clinicians or patients, or finding out about problems associated with products currently available on the market. The goal of this stage is to explore and capture product requirements as seen through the eyes of the users and patients and then use these "user needs" as a target profile to guide the subsequent design phases. The user needs are important because they are the springboard for two subsequent sets of documents, the risk analysis and the specifications for the product. These two sets of documents undergo revision as experience is gained with the design and its performance. Commonly, the user needs are considered to be part of the design inputs. The engineering specifications that correspond to those needs may be viewed by some companies as an "input" to the design and thus put into the "Design Inputs" as well. Alternatively, they can be viewed as the first level of "Design Output" and be put in the design output file.

The design inputs establish the requirements for work that will take place to design a prototype device, which is the final "implementation" of the device to be

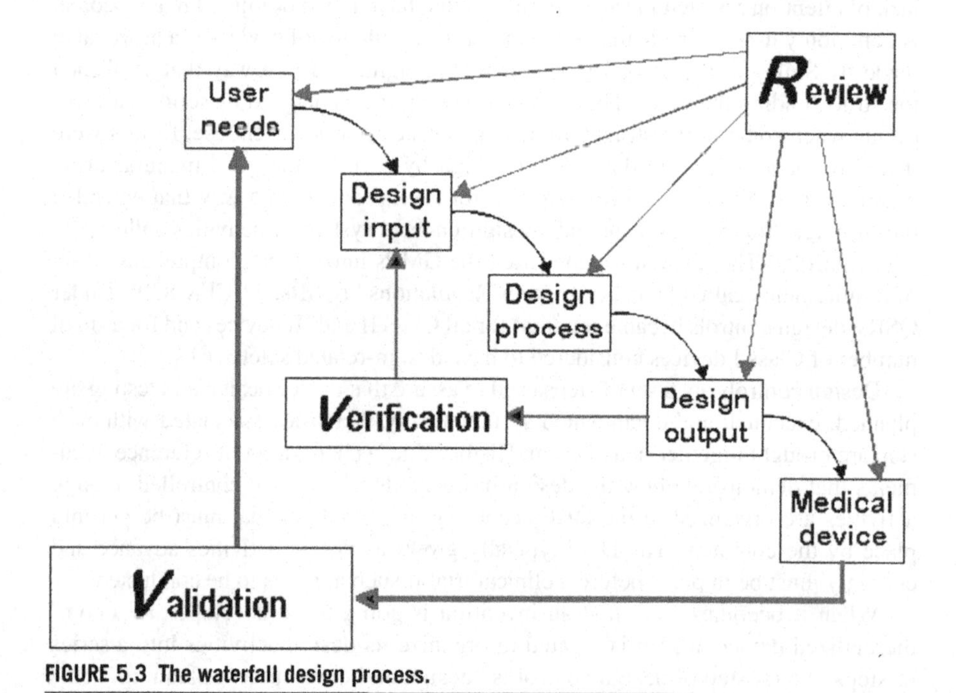

FIGURE 5.3 The waterfall design process.

Source: Design control Guidance for Medical Device Manufacturers, FDA, 1997.

marketed. The design output file grows as it accumulates engineering drawings, work instructions, and other information needed to make the device. The design output should also identify the features of the device that must be tested to verify that the prototype satisfies all of the specifications that were originally set for the device. Thus, the "Design Output" part of the DHF will typically be quite lengthy.

Once a design is advancing to a near-final stage, testing activities must assure that the device is safe and meets all of its specifications. The nature of these tests will vary depending on the device in question. Standards, testing protocols, and guidance documents will all be consulted to assure that the testing is carried out according to best industry practices. These tests may include biocompatibility, mechanical, and electrical tests, for example. The company may also hold an early "pre-IDE" meeting with the FDA in order to be sure that the ideas for testing make sense and that questions about the design and testing are clarified before expensive activities take place unnecessarily. The reports from these tests are gathered to produce a set of documents designated as the "Design Verification." They are organized to provide a compelling argument that all of the specifications for the product were satisfied by the design output.

Once the design output has matured into a prototype, design controls call for a second stage of testing, called "design validation." At this stage, the device is tested to assure that it really does meet the needs that were identified at the beginning of the design phase. In some cases, the validation testing may take the form of clinical trials. In others, it can take the form of human factors or usability testing. Sometimes, it might involve something as straightforward as tests with patients to make sure that they can read and understand the Instructions for Use.

A key element of design controls is "Design Review". At each stage of the design control process, reviews must be conducted by a group of individuals organized into a "design review board" that will critique the work that has been done to satisfy each stage. Minutes of the meetings are recorded and the successful outcome of each stage is documented and signed by responsible individuals capable of arm's length assessment. This series of reviews helps to assure the FDA that the design process is "in control." Although each step of the design control waterfall is typically characterized as an orderly progression, iterative loops are common as development proceeds. For example, a test may show that a particular device component fails its testing, and a new component or even a new design must be developed that will force a return to the Design Input stage.

Two other elements of the design control process remain after a final design is approved. "Design Transfer" consists of the series of activities and meetings to assure that the prototype design can be successfully manufactured in large quantity. Activities can involve scale-up and revision of manufacturing steps to assure that the product can be made reliably and consistently. "Design Change" recognizes that changes to the product will occur over time and identifies how minor and major changes will be defined, and how these changes will be fed back into the design control system. Those changes that could affect the safety or performance of the device, perhaps by interfering with another function or component, must be scrutinized carefully.

Often a change that may be useful for one aspect of product performance can have unexpected consequences for another aspect. Thus, design changes cannot be made without assuring that retesting takes place and the DHF is updated.

5.4 ASSURING COMPLIANT MANUFACTURING

As devices advance through the design phase, developers must prepare for the eventual production of the device in large quantities. The ability to produce the devices consistently and efficiently is a significant undertaking and at least two aspects of this work will have regulatory implications. First, the device must be made in compliance with quality systems regulations (QSRs). Second, a decision must be made about where the manufacturing will be done, whether by the company itself, a contract manufacturing organization, or another supplier or vendor.

5.4.1 ASSURING COMPLIANCE WITH QSRS

Regulators expect that medical devices will be manufactured under a rigorous system of quality assurance before commercialization, unless that requirement is waived. In the United States, these rules are considered as QSRs that are spelled out in 21CFR820. The regulations include rules that govern the control of documents and manufacturing processes, but also extend throughout the life cycle of the product from design, as described above, to distribution and servicing. For large companies with many devices on the market, the systems required for assuring quality are generally well understood and in place. However, those who invent novel devices may not be familiar with the increasing levels of discipline, documentation, and validation that QSRs will require. Thus, new companies may find that the activities needed to put quality systems in place can be both demanding and expensive.

An important aspect of quality systems is its emphasis on risk management. Because every device can be different, it is not possible to write guidance or regulations that address all of the many concerns that might need to be considered in assuring quality. Thus, every device should be supported by risk management activities that begin as soon as possible in the design process and continue throughout the life cycle of the product. The standard that is generally followed for device risk management, ISO 14971, characterizes the risk management process as a series of steps that begin with risk assessment, a stage in which various risks are analyzed and prioritized. The full array of risks must then be evaluated to determine whether the overall level of risk is acceptable given the benefits that the device may provide. The risks inherent to treat a serious condition with no other options may be tolerated when the same level of risk would be unacceptable for a device for a less serious disorder with other treatments. Even if the risks are judged to be acceptable, the risk management standard commonly used in the industry, ISO 14971, drives companies to control risks to the greatest extent possible by considering options such as redesign or the addition of redundant features or alarms.

Early in the design process, many of the assumptions about risk are guesses, but these guesses become validated or modified as the device gains experience in the marketplace. Then, complaints and adverse events associated with the product and with similar competitive products give the quantitative information needed to re-evaluate the risk analysis and suggest management strategies when change is needed.

5.4.2 DEVELOPING CONTRACTS WITH CMOS AND VENDORS

Small companies making products for the first time traditionally rely on the use of a contract manufacturing organization (CMO) for some or all of their manufacturing. However, even large companies often outsource manufacturing in whole or in part and most companies rely on vendors for particular elements of the product, such as printed circuit boards, packaging, or subassemblies. Often, the regulatory implications of such contractual arrangements are not well recognized. Thus, companies may fail to clarify the obligations of vendors, such as the need to adhere to quality systems or communicate certain types of problems with the sponsor as part of their contractual responsibilities. This becomes particularly important if a whole device is outsourced or if a critical component is being purchased rather than produced. Further, if the manufacturing is done off-shore, the vendor may not appreciate the importance of meeting certain quality standards that are based on the country in which the product will be sold, not manufactured.

Whatever the level of outsourcing, the company that holds the marketing approval is responsible for assuring that the product ultimately meets the requirements of the governmental agency in the country in which the product is sold. Expectations of vendors that are spelled out in the contracts must include the requirements for quality metrics and what will be done if the product or part fails to meet the identified specifications. This can be challenging because not all aspects of the quality requirements are harmonized internationally. Some vendors may be accustomed to quality systems in a country whose standards are different. Further, multiple inspections can be anticipated from different nationally affiliated regulatory agencies or third-party inspectors if a product is to be sold globally. Thus, the implementation of appropriate quality systems must go hand-in-hand with decisions about the countries in which the product will be launched. Currently regulatory agencies in some constituencies are working to develop a **Medical Device Single Audit Program**, called **MDSAP**, in which an audit of the quality systems by inspectors from one member of the Program will be accepted by multiple countries.

5.5 CONDUCTING CLINICAL TRIALS

Most medical devices entering the market today do not have to be tested in humans through clinical trials. Clinical trials are, however, needed for most Class III devices and for a small proportion of Class II devices that can only be judged as safe and efficacious with this additional level of evaluation. Companies may also want to perform clinical studies for other reasons, for example, to prove comparative effectiveness in order to support or strengthen its marketing or reimbursement position, or to guide design of usability or customer acceptability.

Typically, a clinical trial is a structured interventional study based on a detailed protocol. In such a trial, the company sponsors the trial by paying for it and working with the regulatory agency to assure that the trial is approved through a submission called an Investigational Device Exemption (IDE). The use of the term, *exemption*, communicates the fact that the device will be exempted from the usual marketing permissions and some of their associated regulatory requirements when it is used experimentally in a patient. The sponsor then obtains the services of one or more principal investigators (PIs) who conduct the trial at a clinical site. The PI is responsible for logistics such as the recruitment of appropriate subjects, the approval and use of informed consent forms and advertising materials, and the ongoing interactions with the Investigational Review Board (IRB) that will oversee the safety and ethics of the work at that institution. In some trials where a device does not pose a significant risk to patients (nonsignificant risk device), the approval of the IRB is sufficient before the trial begins. The trial will be subject only to abbreviated IDE requirements like labeling the device as an investigational device. However, for those trials where the device does pose a significant risk (significant risk device), the sponsor must also obtain IDE approval from the FDA in addition to the IRB approval, before proceeding with the clinical trial. A product with significant risk is considered one that presents a potential for serious risk to patient health. For example, most implanted products would be considered to have significant risk, as would many products that are intended to support or sustain life.

5.5.1 INVESTIGATIONAL DEVICE EXEMPTION

The submissions to an IRB and FDA differ. The IRB submission generally follows a format that is dictated by the hospital or university and thus can vary from institution to institution. For a large multicenter trial, several IRB submissions will need to be made and several versions of the informed consent form may be used in accordance with the specifications of each unique site. The objective of each application is to assure the institution that the protocol will not expose the trial participant to unjustifiable levels of risk. The institution will also want to be sure that subjects understand the risks associated with the trial and any conflicts of interest that might be present in the study team.

An IDE submission to the FDA is required if the device or procedures involved in the trial are considered to have a "significant risk." The IDE submission goes beyond the concerns about patient welfare to evaluate the scientific rationale and protocol design, in order to be sure that the data generated by the study will be valid and useful to support the eventual commercial goals for the device. The IDE typically has three main sections. First, it describes the device design, risk analysis, and testing from an engineering perspective to assure that the risks have been controlled and the device performs according to specifications. Second, it summarizes any animal testing that has been done to show that the device is sufficiently safe for studies in humans and that it has a high probability of benefit. In the case of a device that contacts tissue, it may include biocompatibility testing carried out in accordance with the accepted international standard, ISO 10993-1. The biocompatibility requirements vary according to the duration of exposure and the specific type of tissue contacted by the device. Third, the clinical testing is described. If this is the first time that the device has been used in man, the IDE will often contain only the proposed protocol, with a substantial amount of information explaining where the trial will be carried out, who will do it, and how the company will monitor the trial to be sure that the investigators at the site carry out the study according to the protocol. These studies must be carried out under a system of quality practices that are collectively called good clinical practices (GCPs), designed to assure the integrity of the records and the quality of the work.

The regulatory management of clinical trials can have several stages. The FDA recognizes that more than one device trial may take place, from early interventions with only a few participants to later stage trials with more rigorous protocols and many hundreds of patients. In the past, device trials were divided into two types, feasibility trials and pivotal trials. A feasibility trial is designed to assess patient safety and the usability of the device, typically in a small number of patients at a single site. The information gained at this stage could be used to justify one or more follow-on "pivotal" clinical trials. Results from pivotal trials provide the main body of data that regulators will use to evaluate whether the product is sufficiently safe and efficacious for approval or clearance. In these trials, companies must choose carefully the outcome measures that will produce a statistical level of confidence to support the intended use and safety of the device.

When a company is planning to conduct clinical trials, it will generally schedule a pre-IDE meeting with the regulators. At this meeting, the concerns of the company are discussed and the regulators give feedback about the work done to date and the clinical strategy. These meetings are important to assure that the company does not make unwarranted assumptions or inappropriate projected claims and that it does not design a trial that will be seen to be inadequate when the submission is reviewed. After the company submits its IDE, the FDA has 30 days in which to decide if the study can go forward. A response letter that identifies areas of concern is then sent to the sponsor. In some cases, deficiencies identified by the regulators are not sufficient to prevent the launch of the trial, but in others the trial will be put onto clinical hold until the sponsor answers the questions or makes the adjustments seen as necessary by the reviewers. The FDA also has the option of putting the trial on clinical hold after it has begun if safety issues or problems with trial conduct occur.

When an IDE is submitted for the first time, it does not have a number. A number is assigned to that IDE by the FDA, and this number follows the product throughout the course of its clinical development. Thus, more than one clinical trial may be carried out under the same number, as the clinical program matures and more trials are conducted. The addition of new information to the file is considered to be a "supplement" to the file. Each supplement is numbered sequentially, regardless of the nature of the added information. Several types of information will over time become part of this file. First, the FDA requires an annual report that summarizes the state of the trial and any adverse events or changes that have taken place over the year. Any significant change in the protocol or the information on which it depended for approval will also have to be reported to the FDA as separate supplements throughout the year. Such reports may include, for example, the addition of new procedures or outcome measures, the addition of a new protocol, any serious unexpected adverse events, or any new preclinical data or manufacturing changes. Once a trial is over, the sponsor also must submit a final report to the FDA within 6 months and this will be added to the file. The contents of an IDE are considered to be proprietary and are not shared publicly, but the correspondence between the FDA and the sponsor to indicate the approval status is available to the public through the FDA web site.

A few regulatory changes and requirements have been added recently to the IDE process. One notable change has been the addition of a category of clinical trials, called "early feasibility studies," at the beginning of the development phase. These studies allow for a company to conduct a small trial with an innovative device in order to understand certain aspects of function before launching a larger trial. This information might be needed to inform the development of later larger trials or to gain insights into technological issues that might be improved in the design phase. The early-feasibility and feasibility trials are not usually shared publicly. However, when pivotal trials commence key information about the trial must be published on a US-based registry of clinical trials called clinicaltrials.gov so that patients who might be appropriate subjects for the trial can find details about trials in which they might want to participate. Once the trial is over, results of the trial must also be posted on this website within 12 months, unless permission is requested and obtained from the FDA to delay publication until the product is approved.

5.6 GAINING REGULATORY APPROVAL

Perhaps the most significant milestone in the lifecycle of a regulated device is its market registration. The path to market registration, outlined for various device Classes above, should result from a planned strategy in which engineering activities are orchestrated with necessary preclinical and clinical testing. An increasing emphasis has been placed by FDA on establishing an early partnership in which meetings are held and consultations are suggested to clarify the requirements for a specific product. Typically, the FDA encourages that sponsors meet with the reviewing division associated with the product, which may be conducted through teleconference, videoconference, or face-to-face meetings. FDA has also identified what is called a Q-submission program to track and respond to requests for feedback, including requests for certain types of formal meetings, including "Determination," "Agreement," and "100-day" meetings, that can be set up by written request (Fig. 5.4).

A Determination Meeting is designed to assist the development of Class III devices, so that the company can understand what the FDA views to be the important evidence needed to demonstrate that the device is effective for its intended use. An

Q-Sub Type	Meeting as method of feedback?	Timeframe for meeting/teleconference (from receipt of submission)
Pre-submission*	Upon request	75–90 days**
Informational meeting	Yes	90 days
Study risk determination	No	N/A
Agreement meeting	Yes	30 days or within time frame agreed to with sponsor
Determination meeting	Yes	Date for meeting agreed upon within 30 days of request
Submission issue meeting	Yes	21 days
Day 100 meeting	Yes	100 days (from PMA filing date)

*As defined in MDUFA III Commitment Letter.
**21 days for urgent public health issues (see Section III.A.6.).

FIGURE 5.4 Types of meetings used to request feedback on development programs from FDA.

Source, FDA, 2014.

Agreement meeting, also typically used for Class III devices or Class II implants, is intended to help the company and the FDA to reach agreement on the important elements of the development plan and on the design of the clinical protocol. The 100-day meeting is designed to discuss the status of a PMA application that has already been submitted. Whatever the type of meeting, FDA expects that the company will prepare and send relevant materials to base the discussion, including information about the proposed agenda and questions to be addressed, information about the individuals who will attend the meeting and where relevant, background technical information about the device and its current testing plan. Usually the more formal meetings have timelines for the FDA to respond to the request for the meeting and will have specific guidance about how long in advance such a meeting will have to be set up.

5.6.1 POSTMARKET MANAGEMENT

Once a device is placed on the market, regulatory activities must continue to ensure regular communication regarding changes in product status and production methods to the FDA. In addition, the company must conduct activities related to problems that are identified when the product is in the field, including adverse events, "field actions" and complaints.

5.6.1.1 Reporting and acting on adverse events and complaints

When healthcare professionals or patients encounter problems with a medical device, they often communicate that problem or complaint to someone in the company. The company is obliged to document and investigate each event and then to determine if the adverse event should be reported to the FDA. An adverse event is reportable if the evidence reasonably suggests that a death or serious injury can be attributed to a specific device. An adverse event that is judged as serious is an event that is life threatening, permanently impairs the body, or requires that medical or surgical treatment to prevent permanent injury. It must be reported to the FDA within 30 calendar days after that event first becomes known to anyone in the company, whether it occurs in the United States or another jurisdiction. In addition, malfunctions of long-term implants or life-supporting or -sustaining devices that could "cause or contribute to a serious adverse event if it were to recur" must also be reported in 30 days. Soon, these rules may change to allow Class I and some types of lower risk Class II products to report problems as a group quarterly rather than individually, but those regulations are still in development.

If the reported problem "necessitates remedial action to prevent an unrealistic risk of substantial harm" to the public, FDA must be notified within 5 working days after the company becomes aware of the problem. In addition, follow-up reports must be submitted if new information becomes available related to a previous report. Adverse events and malfunctions are transmitted to FDA using Medwatch Form 3500A. The reports are eventually posted in the Manufacturer and User Facility Device Experience (MAUDE) database that is publicly available on the FDA's web site.

Companies are not the only source of information about adverse events and malfunctions. That information can also come from patients or their families, healthcare professionals, or healthcare facilities. Healthcare facilities are obligated to report on their adverse events each year. All of these reports allow FDA to identify sentinel events or trends that might signal problems, such as design deficiencies, counterfeit products, or manufacturing problems associated with a particular product.

5.6.1.2 Field corrections and recalls

A problem with a product may be sufficiently serious that some action must be taken to withdraw it from the field. Recalls are assigned to one of three classes according to the seriousness of the risk. Unlike device classification that places the lowest risk products in Class I, Class I recalls are those with the highest risk that could pose serious health risks or deaths. Class II recalls are product problems that could cause temporary or medically reversible adverse effects, with only a remote chance of serious adverse events. Class III recalls are the least serious and include violative situations that are unlikely to harm patients. Most recalls of devices are initiated voluntarily by the company, but FDA has the authority to force a recall if the device appears to pose a threat to public health and the company is uncooperative. When a company orders a Class I or II recall, it must inform the FDA in 10 working days; if it orders a Class III recall, such notification is not technically needed but FDA prefers to be notified so that it can independently assess the seriousness of the recall. The information that is transmitted to the FDA should detail the types and batches that are recalled as well as information about how and who will be receiving information about the recall and how the company will measure the success of the recall. It will also explain the actions that the company has taken or will take to correct the issue. In all recalls, careful records must be kept.

A problem that has previously been faced by manufacturers attempting a recall has been the difficulty in tracking the ultimate destinations of products in the field. Tracking medical devices has been of sufficient concern to spur laws that require a unique barcode or a two-dimensional code to be placed on medical devices. This code allows the device to be followed through its manufacture, distribution, and use. Originally intended to be in place by 2016 for all medical products, its full implementation has been delayed so that healthcare facilities have time to put in place the necessary equipment and systems to track the devices. Upon complete implementation, the label of most devices will include a unique device identifier (UDI) in human- and machine-readable form. Device labelers are then required to submit a range of information about each device to FDA's Global Unique Device Identification Database (GUDID). This effort has required significant changes in labeling, documentation, and supply chain structure for both manufacturers and healthcare facilities.

5.6.1.3 Complaints

Companies will receive complaints about their products from many sources, including patients, healthcare professionals, and others in the distribution chain. Even when complaints about devices are not serious enough to warrant a medical device report or recall, they still must be handled carefully. Standard operating procedures must be in place so that each complaint is investigated and acted upon appropriately. Records must be kept to identify how the complaints were dispositioned. Complaints often require corrective and preventive actions (CAPA). Failures to initiate or complete CAPAs following complaints or adverse event reports are among the most cited quality violations when companies are inspected by FDA.

5.7 MEDICAL DEVICE REGULATION IN OTHER COUNTRIES

5.7.1 EUROPE

The European system for medical device registration differs in several respects from that in the United States. In Europe, medical devices can be sold in any country without review by a governmental agency if they bear a "CE (Conformité Européene) mark," by which a company declares its compliance with essential requirements that are specified by certain product directives or regulations. The company does this by working not with the governmental bodies, but rather with a "notified body," an organization that has been accredited by one of the European countries to assure that the product meets the required standards for safety and performance. This activity is described in Europe as "conformity assessment." The notified body is paid by the company to assure that its products have been designed and tested sufficiently. The notified body also inspects the company to be sure that its clinical trials, if any, and its quality systems are in compliance with relevant requirements and standards. If the notified body believes that the company is not complying, it will not allow the company to affix the CE mark. Further, the notified body must inform its accrediting government of significant problems with the product or manufacturer if these cannot be managed effectively. When the company has completed all of the necessary steps to satisfy the device requirements, the notified body assists the company to declare its compliance with the relevant rules and then to put the CE mark on the device.

In other respects, however, the process that is needed to assure compliance is similar to that in the United States. As in the United States, the devices are categorized in a risk-based system, but in that system are four classes, I, IIA, IIB, and III. Class III products have the highest risk. The method to identify the classification of a device is relatively straightforward compared to that in the United States. Rather than relying on a product-specific regulation or predicate, each device stands on its own. Its class is determined by safety and performance characteristics that are matched to a checklist provided in a Directive. As in the US system, the requirements for each device class differ and become more rigorous for higher Classes. Thus, the technical files that are needed in the United States as part of a PMA submission, for example, are similar to the materials needed for what is called a "design dossier" in Europe.

The market registration of a medical device in Europe depends on the ability to satisfy requirements that are detailed in "Directives." At the time of writing, medical products are governed principally by three Directives: the Active Medical Devices Directive 90/385/EE (AMDD), the Medical Device Directives 93/42/EEC (MDD), and the In Vitro Diagnostic Device Directive 98/79/EC (IVDD). This framework will be replaced in the near future by two Regulations, one for medical devices and one for in vitro diagnostics. These revisions will introduce a number of changes in the rules but relatively little change in the overall approach to device management. The Directives, and eventually the Regulations, spell out expectations regarding the requirements for design and testing, the quality system expectations, and the documentation and inspection requirements to gain and keep a CE mark.

Compliance in the EU from a quality perspective depends on satisfying the current version of the international quality standard, ISO 13485: Medical devices—Quality management Systems—Requirements for Regulatory Purposes. The ISO document is organized in a way that appears on first examination to differ from the US regulations for quality systems, but both documents have similar expectations regarding the activities required to implement an effective quality system. Both require that manufacturing will be conducted using similar quality approaches, supported by effective risk management and CAPA programs. They also expect that the approach to quality issues will be risk-based. Activities for Class I products require rather less intensity than higher-risk products. Class I products can even avoid the oversight of a notified body and can self-certify to simpler requirements that are laid out for such products in the Annexes to the ISO document.

The European rules, like the US rules, also require that the quality system be audited regularly while on the market to assure that it continues to meet the specified standards. These inspections are carried out routinely by the notified body, which will reissue a certificate of compliance with ISO 13485 at the time of each audit as long as the company continues to comply effectively with the standard. If, however, the governments of one or more countries believe that the company is violating the standards and this violation has not been effectively identified and communicated by the notified body, it may also conduct its own audit.

5.7.2 JAPAN

The regulatory system in Japan appears to blend the approaches taken in the United States and EU. The Japanese system relies on review and approval of devices under their Pharmaceuticals and Medical Devices Act (PMD Act). It relies on a risk-based system with four classes, Classes I, II, III and IV; Class I products are those with lowest risks and Class IV are those with the highest risks. The classification determines the pathway that will be taken for registering the device. Class I devices can file a marketing submission called a "Todokede" that notifies the Pharmaceutical and Medical devices Agency (PMDA) about the placement of the device on the market, without an expectation that a submission will be reviewed. Class II devices and a few Class III devices with a specific certification standard have a certification process that depends on compliance with relevant standards. The application for approval (the "Ninsho") will be reviewed by a Registered Certification Body, rather like a Notified Body in the EU. That Body will then make a recommendation to the Ministry of Health, Labour and Welfare (MHLW). Other devices in the Class II–IV categories will undergo a preapproval review by the PMDA that will make an approval recommendation to the MHLW. The submission documented to support this review is called a "Shonin" and is in most respects similar to the PMA submitted in the United States.

Like other countries, Japan has an expectation that device will be manufactured under strict quality systems compliant with their own regulations in MHLW Ordinance 169; these expectations are similar to those of the United States and Europe. Japan also requires companies to register design and manufacturing sites. If the company does not have an office in Japan, it will need an in-country representative to be signatory to the submissions and a Marketing Authorization Holder (MAH) license.

5.7.3 CHINA

Medical devices in China are regulated by the Chinese Food and Drug Administration (CFDA), with the support of State and municipal FDAs that are located in provinces and major cities. Devices are placed in three Classes, I, II, and III, that are based on risk. Unlike the other countries, however, China places many more devices into the riskiest category, Class III. Registration processes differ for products from domestic versus international companies. Domestic products in Class I are typically reviewed locally by a municipal FDA, Class II devices by a provincial FDA, and only Class III products are reviewed by the CFDA. In contrast, products of foreign manufacturers will be reviewed by the CFDA regardless of class. The review process in China requires not only the submission of a detailed dossier but also of samples that can be tested by the government to assure that they meet standards that may be specific to China. Chinese authorities also want to see evidence of the approval status of the product elsewhere, including a CE certificate for the EU or a 510(k) clearance letter or PMA approval from the United States.

China also has its own regulations, described in Order No. 10, for quality systems that appear to blend the rules of both the United States and the EU into a very detailed description of quality requirements. The CFDA expects that foreign manufacturers will have agents in China to handle various legal and regulatory obligations. These will include a registration agent, a legal agent, and an after-sales agent, to handle registrations, adverse events/field actions, and service/maintenance support, respectively.

5.7.4 CANADA

In Canada, devices are grouped into four risk-based classes as they are in Europe, but the Classes are numbered I–IV in order of increasing risk. Companies wishing to market in Canada must obtain an establishment license from Health Canada that includes confirmation that the company will fulfill legal requirements with respect to such activities as record-keeping and adverse event reporting. Class I devices do not undergo regulatory review. Each of the other classes have their own review requirements that increase in depth and detail for classes with progressively higher risk. A detailed map of requirements for each class can be identified from a matrix on the Health Canada website. The review is conducted by the Medical Devices Bureau in Health Canada. Devices in Classes II–IV also must have quality systems in place that are compliant with ISO 13485, but Canada has its own very similar version of ISO 13485, that must be audited by a third-party auditor accredited by the Standards Council of Canada and recognized by the Canadian Medical Devices Conformity Assessment Scope (CMDCAS).

5.8 IN VITRO DIAGNOSTICS—A SPECIAL TYPE OF MEDICAL DEVICE

In vitro diagnostic products (IVDs) are a special type of medical device, defined as

> *...reagents, instruments, and systems intended for use in diagnosis of disease or other conditions, including a determination of the state of health, in order to cure, mitigate, treat, or prevent disease or its sequelae. Such products are intended for use in the collection, preparation, and examination of specimens taken from the human body."*

As with other types of medical devices, the FDA further subclassifies IVD products into three classes, Class I, II, or III, according to the level of regulatory control it feels to be necessary to assure safety and effectiveness. The classification of an IVD determines the premarket path in the same way it does for other medical devices.

A unique feature of an IVD is its reliance on reagents, which are central to the nature of the test(s) it performs. There are two types of IVD reagents—General Purpose Reagents (GPRs) and Analyte Specific Reagents (ASRs).

5.8.1 GENERAL PURPOSE REAGENTS

A general purpose reagent is a chemical reagent used in a laboratory "to collect, prepare, and examine specimens from the human body for diagnostic purposes, and that is not labeled or otherwise intended for specific diagnostic application" (21 CFR 864.4010). General purpose reagents can be used alone or together with other GPRs or ASRs as a part of a diagnostic procedure or an IVD test system. The FDA classifies general purpose reagents as Class I medical devices.

5.8.2 ANALYTE-SPECIFIC REAGENTS

Analyte-specific reagents (ASRs) are:

> *... antibodies, both polyclonal and monoclonal, specific receptor proteins, ligands, nucleic acid sequences, and similar reagents which, through specific binding or chemical reaction with substances in a specimen, are intended for use in a diagnostic application for identification and quantification of an individual chemical substance or ligand in biological specimens (21 CFR 864.4020).*

Based on their intended use, ASRs can fall into any of the Class I, II, or III subcategories. Class I ASRs, like other Class I medical devices, are exempt from premarket notification and only require the general controls described above. Class II ASRs require compliance with the special controls and guidance documents. An ASR is classified as a Class III device if the reagent is intended as a component in two kinds of tests. First are tests to diagnose a contagious condition that could easily cause death, but for which prompt and accurate diagnosis offers the opportunity to reduce the public health risk. Examples include human immunodeficiency virus (HIV/AIDS) and tuberculosis (TB). Second are tests used to screen blood donors for conditions, identified by the FDA to assure the safety of blood and blood products. Examples include tests for hepatitis and for identifying blood groups. Class III ASRs require premarket approval (21 CFR 864.4020(b)(3)).

In addition to FDA regulations, IVDs must also comply with the requirements set forth in the Clinical Laboratory Amendments of 1988 (CLIA). These amendments to the Public Health Services Act require certification and oversight of clinical laboratory testing by the respective state, as well as by the Center for Medicare and Medicaid Services (CMS). Laboratories can obtain multiple types of CLIA certificates, based on the kinds of diagnostic tests they conduct. Three federal agencies are responsible for the different aspects associated with the CLIA requirements:

1. FDA—categorizes tests by reviewing the package insert for test instructions and applying a "scorecard" of criteria based on the complexity of the test. They also review requests for a "Waiver by Application" or a CLIA waiver, which is obtained by labs to perform simple tests that have low risk of error, like the glucose meter test. They also develop rules and guidelines for categorizing tests as moderate or high complexity.
2. Center for Medicaid Services (CMS)—issues laboratory certificates, collects user fees, conducts inspections, publishes CLIA rules and regulations, and approves private accreditation organizations, state exemptions, and proficiency tests.
3. Centers for Disease Control (CDC)—provides analysis, research, and technical assistance with development of technical standards and laboratory practice guidelines. They conduct laboratory quality improvement studies, monitor proficiency testing practices, develop and distribute professional information and educational resources, and manage the Clinical Laboratory Improvement Advisory Committee (CLIAC).

5.8.3 PREMARKET SUBMISSIONS FOR IVDs

As described above for medical devices, Class I IVDs must also adhere to general controls, including establishment registration and product listing, but are exempt from a premarket submission. In addition to general controls, Class II and III IVDs require further testing to meet performance standards corresponding to either a premarket notification or a premarket application.

5.8.3.1 IVDs subject to premarket notification—510(k)

The premarket notification submission for an IVD should demonstrate that the IVD is safe and effective by providing comparative evidence that the device performs in the same way as a predicate device. The submission may also include noncomparative analytical studies to establish performance characteristics, such as limit of detection, interference, and cross-reactivity, or a prospective multicenter clinical trial with clinically established endpoints.

5.8.3.2 Premarket approval—PMA

Although the requirements for a PMA application for a class III IVD are similar to those for Class III medical devices, there are some differences. For example, IVDs have different criteria for demonstrating safety and effectiveness than medical devices. Unlike a traditional medical device, the performance of an IVD is not related to the direct contact between the device and patient. Rather, an IVD's performance is related to the downstream impact of the test results to the patient's health and the potential negative consequences if the device were to generate false results. For example, if an IVD test incorrectly identifies a patient infected with HIV as being HIV-negative, there would be serious consequences for that patient and, potentially, for others in the community.

5.8.4 IVDs SUBJECT TO HUMANITARIAN DEVICE EXEMPTION

As with devices in general, an IVD can qualify for status as a Humanitarian Use Device (HUD) if it is to be used to diagnose a disease or condition "that affects or is manifested in not more than 8,000 individuals in the United States per year" (Section 3052 of the 21st Century Cures Act (Pub. L. No. 114-255). Such a product is likely to be used to diagnose or monitor the condition of a rare disease such as a genetic disease, or to support the prescription of a drug for a small subset of patients. For example, an IVD might be used to assure that a small subset of patients has the appropriate cancer marker to support the use of a particular monoclonal antibody. In the future, these types of diagnostics will play a large role in the evolution of personalized medicine.

5.8.5 INVESTIGATIONAL DEVICE EXEMPTION

IVDs posing a significant risk may also be required to undergo clinical trials and will require an IDE if the study poses a significant risk to patients. The rules for IDE submission have been described earlier in this chapter. Clinical evaluation of IVDs that have not been cleared for marketing require an Investigational Plan approved by an Institutional Review Board (IRB). Informed consent must be obtained from all subjects participating in clinical trials and the labeling should state that the IVD is for Investigational Use Only. As with other device trials, the study must be monitored and reports and records maintained. Sponsors of IDEs are also exempt from many aspects of quality systems, but the requirements for design controls (21 CFR 820.30) must still be followed.

5.8.6 LABELING REQUIREMENTS FOR IVDs

The labeling requirements for containers, package inserts, and outer packaging of IVDs are similar to those of other medical devices. The requirements for labeling of the immediate container are listed in 21CFR 809.10(a). An example for an IVD label on the container is given below: (Fig. 5.5)

The label requirements for inserts and outer packaging are listed in 21CFR809.10(b). Inserts will also contain an explanation of the procedure for calculating the unknown, including the definition of each component of the formula, a sample calculation, and the number of significant figures appropriate for the answer. It should define limitations of the procedure, as well as its expected range of values and specific performance characteristics such as accuracy, specificity, precision, and sensitivity. A bibliography is included. These parts of the labeling should also identify the name and place of business of the manufacturer, packer, or distributor and date of issuance of the last labeling revision by the firm.

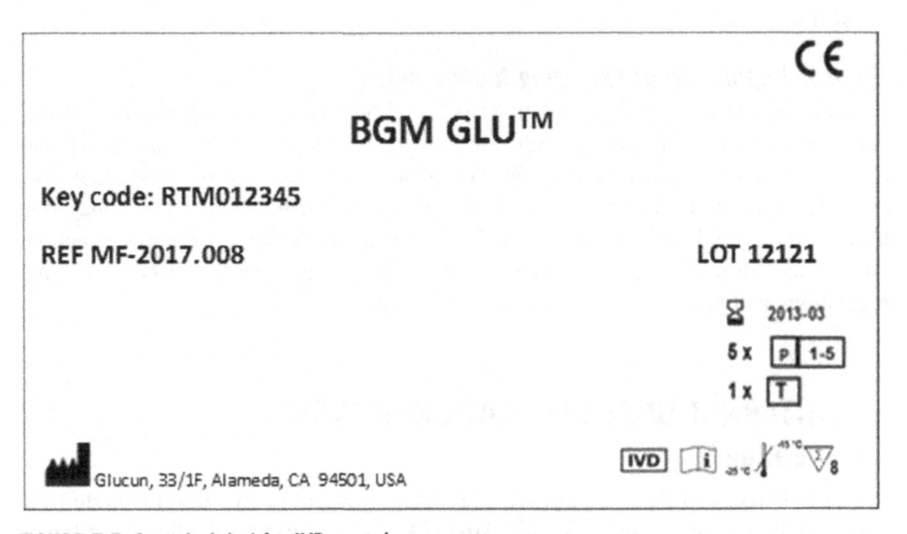

FIGURE 5.5 Sample label for IVD containers.

5.8.7 POSTMARKET REQUIREMENTS FOR IVDs

To ensure that IVDs continue to be safe and effective after they are placed on the market, manufacturers as well as other firms involved in the distribution of devices must follow certain requirements and regulations as described for medical devices in general. These include device-tracking systems, reporting of device malfunctions, serious injuries or deaths, registration of establishments where devices are produced or distributed, and listing of the products that are manufactured and/or distributed. Postmarket requirements also include surveillance studies required under Section 522 of the Act, as well as any other postapproval studies required by FDA at the time of approval of a PMA, HDE or Product Development Protocol (PDP) application.

5.8.8 CHALLENGES FACED BY MANUFACTURERS

The current challenges faced by IVD manufacturers are mainly caused by the uncertainties associated with the changing FDA regulations. Below are some of the key challenges.

5.8.8.1 Regulation of laboratory developed tests

A laboratory developed test (LDT) refers to a type of in vitro diagnostic test that is designed, manufactured, and used within a single laboratory. Over the years, however, LDTs have become more complex, posing higher risk, and some are being distributed and used nationwide. There is a call for increased oversight by the FDA that prompted the Agency to issue a draft guidance on LDT regulatory framework in September 2014. However, the rules and degree of future oversight are still uncertain.

5.8.8.2 Regulation of emerging technologies

FDA has always regulated diagnostic tests based on immunoassay principles. However, with advances in science and technology driving rapid innovation, the regulators will need to keep up with the developments in science and methodologies and consider safety and efficacy of new products. Moreover, innovative IVDs may not have existing predicate devices and will need to pursue de novo pathways to get on the market. More guidance is needed to help companies understand how existing regulations will be applied to the newer technologies.

5.9 INTERNATIONAL REGULATION OF IVDs

5.9.1 EUROPE

The IVD Directive (98/79/EC) provides regulatory requirements that facilitate free trade within the European Economic Area (EEA). The IVDD specifically addresses the safety, quality, and performance of IVDs. The manufacturer of the IVD is responsible for ensuring that their products comply with the Essential Requirements of the Directive before affixing the CE mrk and legally gaining free trade access within the EEA. Per the directive, IVDs are classified into four categories based on the level of risk associated with them. They are:

General IVDs: These include all IVDs other than those covered by Annex II and IVDs for self-testing. Examples of general IVDs include tests for hormones, cardiac markers, and hematology and clinical chemistry tests. For a general IVD, the manufacturer "self-declares" conformity with the relevant essential requirements of the Directive and ensures that the device fulfills the applicable obligations described in Annex III. These devices do not require that the company work with a notified body.

IVDs for self-testing: These include the IVDs used in self-testing, such as pregnancy and cholesterol home tests. This category excludes self-test devices covered in Annex II. As with the general IVDs, a declaration of conformity is required but this declaration will need to be reviewed by a notified body.

IVDs in Annex II: IVDs in Annex II are subdivided into two lists, Lists A and B. All Annex II IVDs require the involvement of a notified body before the product can be placed on the market. IVDs categorized under List B are judged to have moderate risk. They include: Reagents and reagent products, including related calibration and control material for blood groups anti-Duffy and anti-Kidd, irregular, and anti-erythrocytic antibodies, Rubella and toxoplasmosis, Phenylketonuria, CMV, Chlamydia, HLA tissue groups DR, A and B, PSA, self-test blood glucose measuring, devices and software designed specifically for evaluating the risk of trisomy 21.

IVDs categorized under List A are judged to be at high risk. They include reagent and reagent products, including calibrators and control materials for determining blood groups (ABO; Rhesus (C, c, D, E, e); Anti-Kell) and for the detection, confirmation, and quantification of HIV 1 and 2, HTLV I and II and Hepatitis B, C and D.

As with medical devices, the regulatory framework for in vitro diagnostic medical devices will be changed with the adoption of the new EU Regulation 2017/746 for in vitro diagnostic medical devices. The new Regulation will apply after a transitional period of 5 years (spring 2022) whereupon the existing directive (98/79/EC) will be repealed. The new Regulation should provide a stronger premarket review process for high-risk devices, greater oversight of notified bodies, transparency, emphasis on clinical evidence, and better postmarket surveillance. In addition, a new risk classification system for IVDs will be introduced.

5.9.2 JAPAN

Japanese law governs Japanese Medical Device Nomenclature (JMDN), published in the MHLW Official Journal. Once the applicable JMDN is determined, the device registration route, classification/definition, and applied certification/approval standards associated with that product's name are identified and are automatically applicable. Based on the results of classification and JMDN code confirmation, an IVD can be subject to one of three types of IVD registration procedures, that correspond to the procedures of other medical devices in that class. Class III IVDs (and a few designated Class I/II) IVDs must file a premarket approval application with the PMDA and ultimately obtain approval from the MHLW. Novel IVD technology without any published standards or predicate devices must also go through the PMA process. However, most Class II IVDs obtain certification by filing a premarket certification application (Ninsho) with a Registered Certification Body. Class I IVDs file an abbreviated submission that typically has no expectation of technical assessment.

5.9.3 CHINA

The Chinese Food and Drug Administration (CFDA) classifies IVD reagents into Class I, II, and III based on the product risk. Class I IVD reagents include microbial media and products for sample processing like hemolytic agents, whereas Class II IVD reagents include those used for detection of proteins, hydrocarbons, drugs/drug metabolites, and drug sensitivities. Class III IVD reagents include those used in the detection of human genes, hereditary diseases, narcotics, and toxic drugs for medical use.

If a Class II reagent is used in the diagnosis or therapeutic monitoring of serious conditions like cancer or hereditary diseases, it will need to be registered as a Class III reagent. Likewise, if a reagent is used for the detection of narcotics, psychotropic drugs, or toxic drugs for medical use, including their corresponding metabolites, the reagent will need to be registered as a Class III reagent.

5.9.4 CANADA

Regulation of IVDs by Health Canada is essentially the same as that for medical devices discussed earlier under the Section "Medical Device Regulation in Other Countries." But some differences may be encountered for certain products. For example, blood glucose meters require an "intended use" section in their labeling.

FURTHER READINGS

Becker, K. M., & White, J. J. (2010). *Clinical evaluation of medical devices*. Totowa, NJ: Humana Press.

Danzis, S. D., & Flannery, E. J. (2010). *In vitro diagnostics: The complete regulatory guide*. Washington, DC: FDLI Institute.

De Marco, C. T. (2011). *Medical device design and regulation*. Milwaukee, WI: ASQ Quality Press.

Frank, S. (2003). *A new model for European medical device regulation*. Groningen, the Netherlands: Europa Law.

Kahan, J. S. (2014). *Medical device development: law and regulation*. Needham, PA: Parexel/Barnett.

Ramakrishna, S., Tian, L., Wang, C., Liao, S., & Teo, W. E. *Medical devices: Regulations, standards and practices*. Cambridge: Woodhead Publishing.

CHAPTER

Combination products, borderline products, and companion diagnostics

Shayesteh Fürst-Ladani

SFL Regulatory Affairs & Scientific Communication, Basel, Switzerland

> *FDA expects to receive large numbers of combination products for review as technological advances continue to merge product types and blur the historical lines of separation between FDA's medical product-centers*
>
> **– FDA**

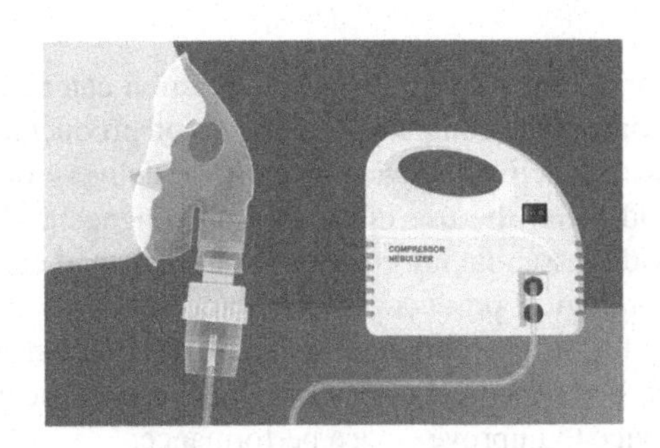

An Overview of FDA Regulated Products. http://dx.doi.org/10.1016/B978-0-12-811155-0.00006-5

CHAPTER OBJECTIVES

After reading this chapter, the reader will be able to:

- define the terms combination product and companion diagnostic from the United States and European regulatory perspectives,
- describe how combination products are designated for regulatory reviews in the United States and Europe,
- explain the United States and European Union (EU) procedures for assigning borderline combination products to a regulatory pathway,
- describe the key stages leading to approval of combination products that are regulated as a drug/biologic or a medical device, and
- summarize how in vitro diagnostic devices are regulated in the United States and EU territories.

6.1 WHAT IS A COMBINATION PRODUCT OR A COMPANION DIAGNOSTIC AND WHY ARE THESE TYPES OF PRODUCTS SO IMPORTANT?

Depending on your area of expertise, the above question can raise several different thoughts. For a pharmaceutical company, a combination product may mean the combination of a drug with a delivery device to improve delivery of the drug product. Alternatively, it could mean the use of an in vitro diagnostic device (companion diagnostic) that provides essential information for the safe and effective use of a corresponding drug or biological product. The combined use of several drugs, such as antiretroviral therapy for the prevention or treatment of HIV, can also be defined as a combination product. For a medical device company, it may mean the addition of a drug to a medical device to improve device performance.

We will define a combination product as a therapeutic or a diagnostic product that combines a drug and/or biological product with a device or in vitro diagnostic (companion diagnostic). The subject matter is important, as the area of combination products permits the development of new and innovative healthcare solutions and continues to show significant growth in the pharmaceutical and medical device sectors (BBC Research, 2015). Advancements in the field of human genomics continue to improve our understanding of diseases at the molecular level. As a result, the emergence of genomic medicine has enabled the development of companion diagnostics that offer targeted treatments for more successful outcomes in patients.

Advances in scientific knowledge and technology are driving growth and innovation for combination products, blurring the historical lines separating traditional drugs, biologics, and medical devices. As a consequence, the regulatory pathways for these products raise unique challenges when compared to drugs, devices, or biological products alone and impact many aspects of product lifecycle management, including preclinical and clinical development, regulatory approval processes, manufacturing and quality control, postmarket surveillance, adverse event reporting, post-approval modifications, promotion, and advertising.

This chapter will examine the various regulations governing product review and regulatory pathways required to gain regulatory approval of combination products, including companion diagnostics.

6.2 CHALLENGES FOR THE REGULATION AND CLASSIFICATION OF COMBINATION PRODUCTS AND COMPANION DIAGNOSTICS

Combination products, including companion diagnostics, are a marriage of different scientific areas involving complex technology. Each part of the product is governed by a different regulatory pathway with different requirements throughout the product development lifecycle. Due to the differences in regulatory requirements and the approval processes of medical devices compared to drugs and biologics, the correct classification of a combination product is a key milestone, and one which is vital to determine early in the development lifecycle.

With rapid advancements in technology come many challenges for regulators. Regulators must adapt quickly to understand the science behind the advancement and regulate new technologies in a consistent manner. They must also provide a level-playing field for manufacturers and ensure the safety of public health, while supporting innovation. In this chapter, we will examine how combination products are regulated in two major regions: the United States and Europe.

6.3 US PERSPECTIVE

Before looking at the definition of a combination product, it is important to understand the individual definitions of the constituent product parts, these being a drug/biologic and a medical device/in vitro diagnostic.

The definition of a drug, as defined in Section 201(g)(1) of the Food, Drug, and Cosmetic Act (FDCA), is

- a substance recognized by an official pharmacopoeia or formulary,
- intended for use in the diagnosis, cure, mitigation, treatment, or prevention of disease,
- intended to affect the structure or any function of the body (other than food), or
- intended for use as a component of a medicine (but not a device or a component part or accessory of a device).

Biological products, as discussed in Chapter 4, are included in this definition and are generally covered by the same laws and regulations. However, differences exist in their manufacturing processes in that drugs are typically manufactured using chemical processes, while biologics are typically manufactured using biological processes.

As discussed in Chapter 5, a medical device is: "An instrument, apparatus, implement, machine, contrivance, implant, in vitro reagent, or other similar or related article, including a component part, or accessory which is:

- recognized in the official National Formulary, or the United States Pharmacopoeia, or any supplement to them,
- intended for use in the diagnosis of disease or other conditions, or in the cure, mitigation, treatment, or prevention of disease, in man or other animals, or
- intended to affect the structure or any function of the body of man or other animals, and
- which does not achieve its primary intended purposes through chemical action within or on the body of man or other animals and
- which is not dependent upon being metabolized for the achievement of any of its primary intended purposes."

Simply put, a medical device acts in a physical manner on the structure–function of the body and does not achieve its primary intended purposes through chemical action.

6.3.1 CLASSIFICATION

In the United States, combination products are defined in 21 CFR 3.2(e) (FDA, 2017a) which includes:

- a product comprising two or more regulated components, that is, drug–device, biologic–device, drug–biologic, or drug–device–biologic, that are physically, chemically, or otherwise combined or mixed and produced as a single entity (examples include drug-eluting stents, antimicrobial catheters, prefilled syringes, or injector pens);
- two or more separate products packaged together in a single package or as a unit comprising drug and device products, device and biological products, or biological and drug products (examples include a surgical kit with a catheter, gloves and antimicrobial wipes, or a drug/biological product packaged with an unfilled injector pen);
- a drug, device, or biological product packaged separately that, according to its investigational plan or proposed labeling, is intended for use only with an approved individually specified drug, device, or biological product (in this case, because both are required to achieve the intended use, indication, or effect, the labeling of the approved product would need to reflect the information corresponding to the associated product).
- any investigational drug, device, or biological product packaged separately that, according to its proposed labeling, is for use only with another individually specified investigational drug, device, or biological product where both are required to achieve the intended use, indication, or effect (examples include a photodynamic therapy drug and light source, a drug and a dedicated but separately provided autoinjector, or a drug and pharmacogenomic test, like an in vitro diagnostic).

6.3.2 DESIGNATION PROCESS

Products are reviewed by the various lead centers at the FDA according to their product classification; this includes the Center for Biologics Evaluation and Research (CBER), the Center for Drug Evaluation and Research (CDER), and the Center for Devices and Radiological Health (CDRH). However, as new combination products emerged, it became challenging to determine which FDA lead center should be conducting the review. As a result, the FDA's Office of Combination Products (OCP), established in 2002, became responsible for assigning a lead center with primary jurisdiction for the review and regulation of a combination product.

Under Section 503(g)(1) of the FDCA, combination products are assigned to an FDA lead center based on the product's "Primary Mode of Action (PMOA)," which is defined as "the single mode of action of a combination product that provides the most important therapeutic action of the combination product." The lead center has oversight responsibility for the review and regulation of the combination product, with an intercenter consultation between the relevant lead centers being available.

Other responsibilities of the OCP include:

- ensuring the premarket review is conducted in a timely and effective manner;
- the coordination of reviews involving more than one lead center;
- ensuring the consistent and appropriate postmarket regulation of combination products;
- the resolution of any disputes regarding the timeliness of combination product review;
- the development of guidance and regulations for combination products to ensure a transparent, consistent, and predictable process.

6.3.3 BORDERLINE CASES—REQUEST FOR DESIGNATION

Borderline cases represent combination products where the associated PMOA cannot be determined with reasonable certainty. In these cases, 21 CFR 3 provides an assignment algorithm to determine which FDA lead center reviews and regulates the combination product.

Lead center assignment may be difficult in cases where:

- the product is in early development, when it is not possible to determine the PMOA
- the product has two (or more) completely different modes of action, neither of which is subordinate to the other.

Fig. 6.1 provides a summary of the FDA lead center assignment algorithm and details the types of questions raised during this process.

In borderline cases, a company may submit a Request for Designation (RFD), as detailed in 21 CFR 3.7. This process requires the company to submit product details, including composition, intended claims, the mode of action, and the company's requested assignment. The FDA has 60 days to review the RFD. If a decision

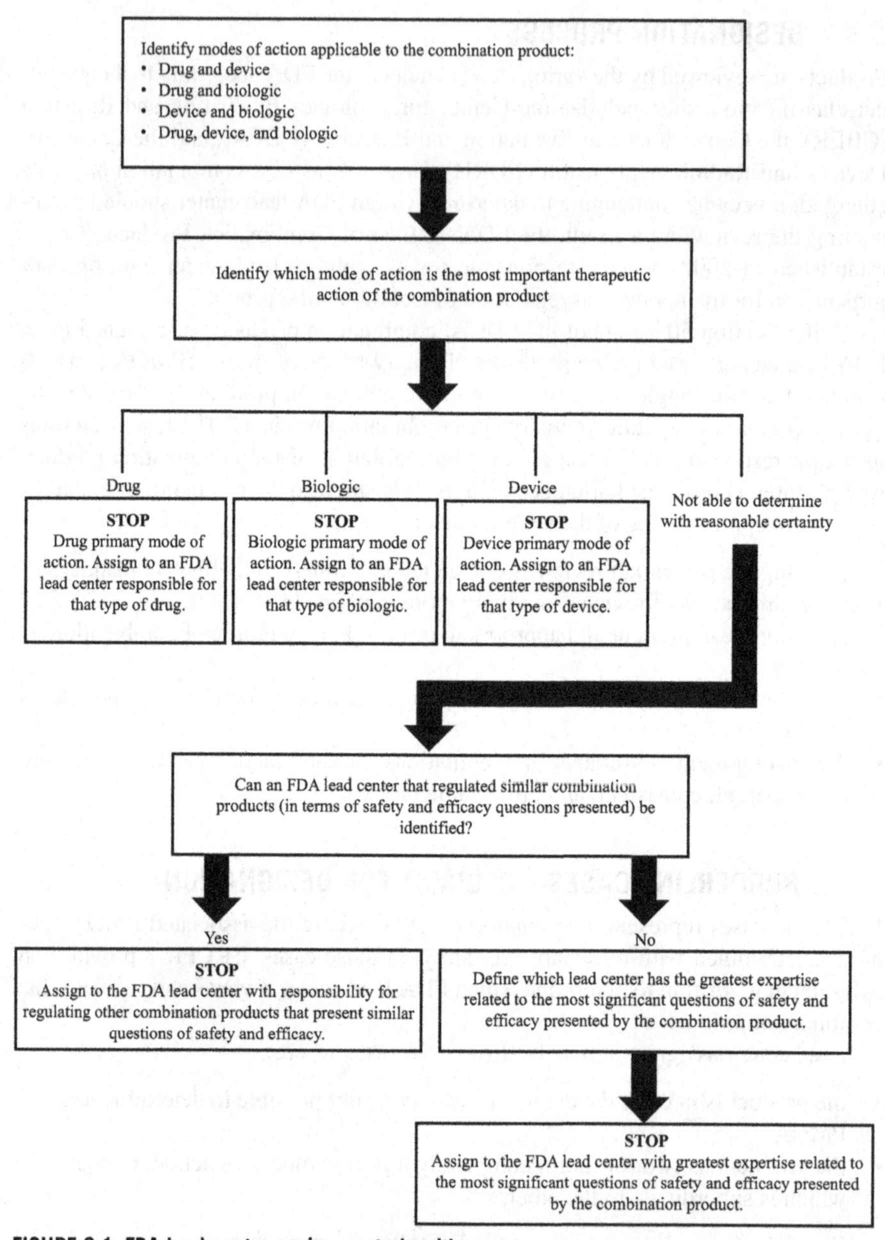

FIGURE 6.1 FDA lead center assignment algorithm.

is not made by Day 60, the company's requested assignment is considered acceptable. In the majority of cases, the FDA provides an RFD decision letter, which details the classification and the assignment of the combination product, identifies the lead

Table 6.1 Summary of Differences Between Drug–Biologic and Device Regulatory Approval

Product Lifecycle Milestones	Drug–Biologic	Device
Preapproval filing	IND	IDE
Regulatory filing	NDA/BLA	510(k)/PMA
Data requirements	CMC pharmacology, non-clinical including toxicology/clinical pharmacology/clinical efficacy and safety	Bench testing, animal data, biocompatibility, national/international standards, clinical data
Clinical studies	Phases, 1, 2, and pivotal (phase 3) studies are usually required	PMA—feasibility and pivotal studied are usually required 510(k)—usually not required
Application review	NDA/BLA—6–10 months	PMA—180–320 Days 510(k)—90 days HDE—75 days
Regulatory Data Exclusivity	Yes	No
User Fees 2016[a]	$2.4M USD NDA/BLA	$261K UDS PMA $5K USD for a 510(k)

[a]*User Fees are revised on an annual basis.* BLA, *Biologic license application;* CMC, *chemistry, manufacturing and controls;* HDE, *humanitarian device exemption;* IDE, *investigational device exemption;* IND, *investigational new drug application;* NDA, *new drug application;* PMA, *premarket approval;* USD, *US dollars.*

center, and provides an overview of the process that was followed to make the decision. If a company disagrees with FDA's decision, it can pursue one of several appeal processes, including a request for reconsideration, as detailed in 21 CFR 3.8, or an internal agency review of the decision, in accordance with 21 CFR 10.75.

To reliably assess the safety and efficacy of a combination product, the assignment of a product to the correct lead center that has the most relevant competence and expertise is essential. Correct lead center assignment also ensures consistent regulation of such products. Additionally, development route, clinical trials, time to market, and associated costs differ depending on the product classification and lead center involved. Table 6.1 provides a summary of the differences between drug/biologic and device regulatory requirements and costs.

6.3.4 REGULATORY APPROVAL PROCESS

If the PMOA of a combination product is the drug or the biologic, the lead center would be either CDER or CBER and the regulatory process would be as described in Chapter 3 (drug) and Chapter 4 (biologic) of this book. If the PMOA of a combination product is the device, the lead center would be CDRH and the regulatory process would be as described in Chapter 5 (medical device) of this book.

As we discussed in Chapter 5, a 510(k) is the submission made to the FDA to demonstrate that the device to be marketed is safe and effective by proving substantial equivalence to a legally marketed US device (predicate device), which is not subject to premarket approval (PMA). A claim of substantial equivalence does not mean the device(s) must be identical. Substantial equivalence is established with respect to: Intended use, design, materials, performance, safety, effectiveness, labeling, biocompatibility, standards, and other applicable characteristics. The 510(k) route is the pathway routinely used for Class I and Class II devices, where there are substantially equivalent devices available. Novel Class II and Class III devices are subject to PMA as discussed in Chapter 5. While the majority of combination products are approved under the PMA route, some, like the drug delivery devices copackaged with a drug, may fall under 510(k).

6.3.5 DATA REQUIREMENTS

Medical devices are subject to the general controls of the FDCA, which are contained in the procedural regulations in Title 21 CFR Part 800–1200. As discussed in Chapter 5, these controls apply to all medical devices and are the baseline requirements necessary for marketing, proper labeling, and monitoring of postmarket performance:

- establishment registration
- medical device listing
- 510(k), unless exempt, or PMA
- investigational device exemption (IDE) for clinical studies
- quality system regulation
- labcling requirements
- medical device reporting.

Detailed guidance on these requirements is available from the FDA web site (FDA, 2017b). The manufacturer must also establish and follow the quality systems regulation before and during the regulatory approval process.

If the PMOA of a combination product is that of a high-risk device, the product would be classified as a Class III medical device and require a PMA. The regulatory review involves intercenter consultations from CDER or CBER to solicit their opinions on the drug or the biologic component of the product. The applicant must receive FDA approval of its PMA application prior to marketing the device. To meet the standard for approval, the PMA application must contain (or include by reference) valid scientific evidence to provide a reasonable assurance of safety and effectiveness of the device when used in accordance with its labeled indication. Therefore, if a combination device includes a drug component, the PMA must also include corresponding drug information, like systemic pharmacology, toxicology, safety, chemistry, manufacturing, and control of the drug. This information is necessary to fully understand the safety profile of the final product. In addition, data from nonclinical and clinical studies are needed to demonstrate the safety of the combination device. The finished combination product should be fully characterized and include engineering studies, biocompatibility evaluation, and animal studies as well as the complete chemistry, manufacturing, and controls (CMC) information, including sterilization, packaging, and shelf life/stability testing.

6.4 EUROPEAN PERSPECTIVE

As described in Chapter 5, in Europe, medical devices are regulated under the medical devices directive 93/42/EC (MDD) as amended by 2007/47/EEC (The Council of the European Communities, 1993). The MDD belongs to new approach directives, in which the company makes a declaration of conformity to the requirements of the applicable directive. This process involves the assessment of conformity of the company's quality management system and technical documentation by a third-party organization called a notified body (NB).

NBs are assigned a product scope and status by the competent authority (CA) in the Member State in which they are located. Lists of NBs, the tasks for which they have been notified and their identification number are freely available on the NANDO (New Approach Notified and Designated Organizations) web site (The European Commission, 2017).

6.4.1 RISK CLASSIFICATION OF MEDICAL DEVICES

Medical devices in Europe are classified by risk to the patient or the user in a manner similar to that followed in the United States. There are five risk classification categories in the MDD. Product risk is classified through assessment of a product's attributes against the 18 classification rules detailed in Annex IX of the MDD.

Risk classifications are:

- Class I medical devices—do not require NB involvement
- Class I medical devices with sterile and/or measuring attribute—require NB involvement for confirmation of sterility and measuring aspects only
- Class IIa medical devices—low-to-medium risk requiring NB assessment
- Class IIb medical devices—medium-risk requiring NB assessment
- Class III medical devices—high-risk classification requiring NB involvement.

6.4.2 COMBINATION PRODUCTS AND THEIR REGULATION IN EUROPE

Although combination product is a widely used term in Europe, it has no associated legal or regulatory definition. The term is applied most commonly when referring to products having both a device and a medicinal product (the term used in Europe for drug) element. These products are currently regulated as either a medical device or a medicinal product based on the "Principal" Mode of Action, comparable to the "Primary" Mode of Action in the United States.

Possible scenarios of where a medicinal product may be combined with a medical device and the regulatory option for that combination, are as follows:

- A medical device that is intended to administer a medicinal product, but does not contain or is packaged with the medicinal product, is regulated as a medical device under the MDD. An empty syringe is an example of such a device.
- A medical device that is placed on the market with an integral medicinal product is regulated as a medicinal product under the medicinal products directive 2001/83/EC (The European Parliament and the Council of the European Union, 2001). An auto injector pen containing insulin as an integral part would be an example of such a product.
- A medical device that incorporates an integral substance (i.e. if used separately can be considered as a medicinal product or human blood derivative, and the action of the medicinal product/blood derivative is ancillary) is regulated as a Class III medical device under Rule 13 of the MDD. Examples of these types of devices include drug eluting coronary stents or a catheter with an antimicrobial coat.
- A medical device that is placed on the market with an integral advanced therapy medicinal product (ATMP) is regulated under ATMP Regulation (EC) No. 1394/2007 (The European Parliament and the Council of the European Union, 2007). Autologous chondrocytes seeded onto collagen membrane to repair cartilage is an example of such a product.

In Europe, a product that contains both a medical device and medicinal product element will be regulated as either a medical device or a medicinal product. The classification depends on a few key factors including:

- The presentation of the product on the market and product characteristics;
- The manufacturer's intended purpose for the product;
- The principal mode of action.

The European definitions for a medical device and a medicinal product are the key starting points when determining the appropriate classification of the combined product. A medical device as defined in Article 1(2)a of Directive 93/42/EEC, as amended states that a "medical device" refers to any instrument, apparatus, appliance, material, or other article, whether used alone or in combination, including the software necessary for its proper application intended by the manufacturer to be used for human beings for the purpose of:

- diagnosis, prevention, monitoring, treatment, or alleviation of disease,
- diagnosis, monitoring, treatment, alleviation of, or compensation for an injury or handicap,
- investigation, replacement, or modification of the anatomy or of a physiological process,
- control of conception,

and which does not achieve its principal intended action in or on the human body by pharmacological, immunological, or metabolic means, but which may be assisted in its function by such means.

Rule 13 of the MDD does allow for the incorporation of a known medicinal product or human blood derivative and states:

> *"All devices incorporating, as an integral part, a substance which, if used separately, can be considered to be a medicinal product, as defined in Article 1 of Directive 2001/83/EC, and which is liable to act on the human body with action ancillary to that of the devices, are in Class III. All devices incorporating, as an integral part, a human blood derivative are in Class III."*

Examples of medical devices that fall under rule 13 of the MDD include drug eluting coronary stents, dressings and plasters incorporating silver, antimicrobial catheters, antibiotic-loaded bone cements, antibacterial-releasing dental restorative materials, and biologic wound care products containing antimicrobial agents.

A medicinal product as defined in Article 1(2) of Directive 2001/83/EC, as amended is:

1. Any substance or combination of substances presented as having properties for treating or preventing disease in human beings or
2. Any substance or combination of substances, which may be used in or administered to human beings either with a view to restoring, correcting, or modifying physiological functions by exerting a pharmacological, immunological, or metabolic action, or to making a medical diagnosis.

As a general rule, medical devices act by physical means, while medicinal products act by pharmacological, metabolic, or immunological means.

It is also worth noting that Article 2(2) of Directive 2001/83/EC, as amended states that:

> *In cases of doubt, where, taking into account all its characteristics, a product may fall within the definition of a "medicinal product" and within the definition of a product covered by other Community legislation the provisions of this Directive shall apply.*

This "in case of doubt" clause is often used in cases where there is insufficient scientific data available to provide clear evidence of the principal mode of action. In these cases, the product would be regulated as a medicinal product.

With technological advancements, especially in the materials science, differentiating the principal mode of action between a medical device (physical) function and medicinal product action has become challenging. Products are increasingly falling into the borderline category and require advice from the working group on borderline and classification. This working group is chaired by the European Commission and is composed of representatives of all EU member states, the EFTA (European Free Trade Commission), and other stakeholders. Each product is evaluated on a case-by-case basis. The working group issues a manual of borderline opinions (The European Commission, 2015) on a biannual basis, which, while not a legally binding document, does provide guidance on a range of borderline products and the consensus of opinion reached by the group.

Another point to note is that, across the EU competent authorities, there are differing opinions in terms of product regulation whereby a product has a physical mode of action but is administered orally. In general, these products are regulated as medicinal products. However, this opinion currently varies across Member States and it is important to be aware of regional differences.

Given the complexity of many of the borderline products and the impact that taking an incorrect regulatory pathway could have on cost and timelines, it is vital to consider the classification of products early on in the development process and, where necessary, discuss the details and regulatory strategy with either an NB or a CA.

6.4.3 APPROVAL PROCESS FOR A MEDICAL DEVICE CONTAINING ANCILLARY DRUG/HUMAN BLOOD DERIVATIVE (CE CERTIFICATION)

In the case where the combination product is considered a medical device, that is, the device contains an ancillary medicinal substance or ancillary human blood derivative under Rule 13, the device must be CE marked before being placed on the market. The conformity assessment routes available for Class III medical devices are summarized in Fig. 6.2.

Most manufactures follow the full quality assurance and design examination route to conformity as detailed in Annex II of the MDD. The alternative conformity assessment route for Class III medical devices is following an Annex III-type examination by an NB with either Annex IV, in which every device/batch is verified by an NB (for nonsterile products only), or Annex V production quality assurance audit by an NB according to ISO 13485:2016 (excluding design). Due to the complex nature of such medical devices, the costs to set-up and validate the testing required to

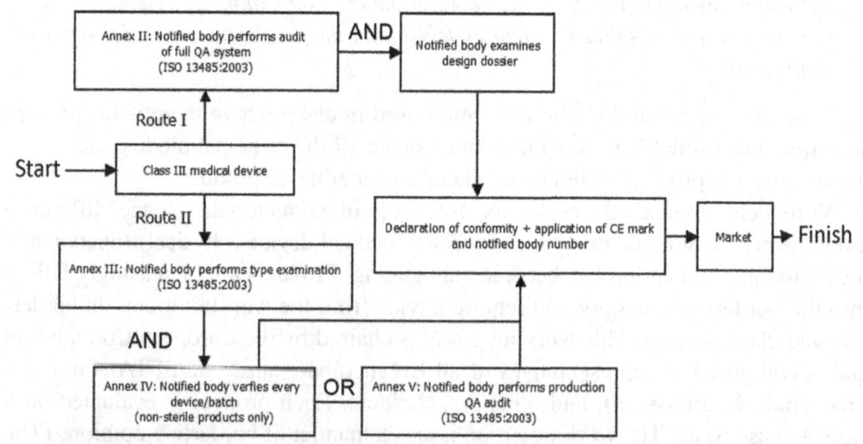

FIGURE 6.2 Class III medical device conformity assessment routes as detailed in Article 11 of the MDD (MHRA, 2008).

conduct type examination testing are extremely high, making this a rarely followed conformity assessment route.

The conformity assessment process is conducted by an NB who must have scope for such assessment. Essential Requirements (ER) 7.4, Annex I of the MDD is applicable to this process. The process requires that, in addition to the assessment of the technical documentation for the device aspects by the NBs technical experts, the "quality, safety and usefulness of the substance must be verified by analogy with the methods specified in Annex I of Directive 2001/83/ EC." What this relates to is that the NB is required to seek a scientific opinion from any CA designated for medicinal products by the member states or the European Medicines Agency (EMA), for human blood derived substances, through Regulation (EC) No 726/2004 (The European Parliament and the Council of the European Union, 2004).

An EMA consultation following Regulation (EC) 726/2004 is also applicable for the regulation and consultation route required for medicinal products developed by biotechnological processes, including:

- recombinant DNA technology;
- controlled expression of gene coding for biologically active proteins in prokaryotes and eukaryotes including transformed mammalian cells;
- hybridoma and monoclonal antibody methods.

A common misconception is that the NB must use the CA in their country for the combination product consultation process. However, there are many factors that are taken into consideration when selecting the appropriate CA, including but not limited to:

- CA's experience of such consultations and resource availability to take on this work
- CA's in-house technical expertise for certain medical device types

The NB is responsible for seeking the opinion from the CA; and prior to the submission of the consultation documentation, the NB must review the manufacturers documentation and verify the usefulness of the ancillary medicinal substance or human blood derivative within the device. This review must include the provision of an opinion on the clinical risk–benefit profile for the incorporation of the substance into the device. Following the receipt and validation of the documentation by the CA or EMA, the consultation process can take up to 210 days with clock stops for questions. In reality, the average time for completion of a consultation is 9 months.

On completion of the assessment, the CA issues a report to the NB detailing either a positive or a negative opinion. For devices containing ancillary human blood derivatives/recombinant DNA technology, a consultation with the EMA is mandatory. If the EMA opinion is negative, the NB must not issue the CE certificate. While the directive does not explicitly state this for consultations conducted with CAs, it would be extremely unusual for a NB to ignore the negative opinion of a competent authority. In all cases, the NB is required to provide the CA/EMA with the conclusions of its assessment and the decision to either issue/not issue the CE certificate.

6.4.4 APPROVAL PROCESS FOR A MEDICINAL PRODUCT WITH AN INTEGRAL DRUG DELIVERY SYSTEM

In the EU, as highlighted earlier, a medicinal product with an integral drug delivery system is regulated as a medicinal product. This combination product follows the medicinal product directive 2001/83/EC, with assessment of the whole product by a CA for medicines, following either a national application, mutual recognition, or centralized procedure. While it is not stated within directive 2001/83/EC, Article 1.3 of the MDD does state that the drug delivery device must conform to the ERs as detailed in Annex I of the MDD.

The level of scrutiny of the device aspects is conducted by the CA with no involvement from a medical device NB. However, as drug delivery devices increase in complexity, the Medical Devices Regulation (MDR) has identified this as a gap in legislation. The MDR seeks to address this issue by calling for a revision to directive 2001/83/EC to allow for NB involvement in assessment when required.

6.5 THE PROPOSED MEDICAL DEVICE REGULATION

The current EU regulatory framework for medical devices consists of the Medical Device Directive (MDD) 93/42/EEC, the Active Implantable Medical Devices (AIMD) Directive 90/385/EEC and the In Vitro Diagnostic Devices Directive (IVDD) 98/79/EC (The Council of the European Communities, 1990, 1993; The European Parliament and the Council of the European Union, 1998). In September 2012, the European Commission proposed a revision of the system, whereby both the MDD and AIMD directives shall be combined under a single regulation for medical devices (MDR), with a separate regulation for IVDs (IVDR). These two new laws shall be introduced as regulations rather than directives, removing the requirement to transpose the text into national legislation at each member state. The regulations were published in May 2017 with a 3-year transition period for the MDR (Regulation 2017/745) and 5-year transition period for the IVDR (Regulation 2017/746) to become fully applicable in 2020 and 2022, respectively (The European Parliament and the Council of the European Union, 2017a, b).

Key common aspects to the new MDR and IVDR include:

- Formation and involvement of new expert bodies. This includes the medical device coordination group (MDCG) which represents Member States, supports the implementation of regulations and makes decisions on the classification of borderline products. EU reference laboratories will also be called upon to provide scientific assistance.
- Introduction of the requirement for a company to appoint a person responsible for regulatory compliance.
- Reinforced rules for clinical evidence.
- Introduction of an additional scrutiny process for high-risk medical devices, active devices intended to deliver a medicinal product and high-risk IVDs. This

process shall involve assessment of the NB conclusion on clinical data, adding an additional step in the conformity assessment process.
- Increased transparency, with information accessible to the public through a central, electronic EU portal.
- Better traceability of devices with unique device identification (UDI) system implementation.

For industry, the greatest impacts of the new regulation are foreseen in the in vitro diagnostic area. Changes will include conversion of IVD classification from a list based on a risk-based approach and provision of a definition for companion diagnostics. The IVDR also increases the requirement for NB involvement in the process of CE certification.

6.6 REGULATION OF COMPANION DIAGNOSTICS

A companion diagnostic is an in vitro diagnostic device used in conjunction with a therapeutic product to identify target patients who would benefit from the treatment. With increasing emphasis on personalized medicine, companion diagnostics are expected to figure more prominently in therapeutics development. We will review how these products are currently regulated in the United States and Europe.

6.6.1 US PERSPECTIVE

An IVD companion diagnostic device is an in vitro diagnostic device that provides essential information for the safe and effective use of a corresponding therapeutic product. The use of an IVD companion diagnostic device with a therapeutic product is stipulated in the instructions for use and labeling of both the diagnostic device and the corresponding therapeutic product. This includes the labeling of any generic equivalents of the therapeutic product.

An IVD companion diagnostic device (FDA, 2014) could be essential for the safe and effective use of a corresponding therapeutic product to:

- identify patients who are most likely to benefit from the therapeutic product;
- identify patients likely to be at increased risk for serious adverse reactions as a result of treatment with the therapeutic product;
- monitor response to treatment with the therapeutic product for the purpose of adjusting treatment (e.g. schedule, dose, discontinuation) to achieve improved safety or effectiveness;
- identify patients in the population for whom the therapeutic product has been adequately studied, and found safe and effective, that is, there is insufficient information about the safety and effectiveness of the therapeutic product in any other population.

The development of products in this area has increased dramatically over recent years. This is particularly due to technological advancements in human genetic testing that can identify patients most likely to benefit from a therapy and, thus, make it possible to personalize medical therapy. Because the results from diagnostic devices are directly linked to determining patient treatment, confidence in device performance to provide accurate results is key. The FDA assesses the safety and effectiveness of both the IVD companion diagnostic device and the therapeutic product through premarket review and clearance or approval.

In general, IVD companion diagnostic devices are in Class III and, therefore, require PMA. Should an IVD companion diagnostic be considered a Class II device, a 510(k) is permitted. The therapeutic product application will be reviewed and approved under Section 505 of the FDCA (drug products) or Section 351 of the Public Health Service Act (biological products) and relevant drug and biological product regulations. FDA reviews of the IVD companion diagnostic device and the drug–biologic product are conducted collaboratively between the relevant FDA offices.

Table 6.2 summarizes the various scenarios that can arise for IVD companion diagnostic devices and therapeutic products and the possible outcomes in each scenario.

Table 6.2 IVD Companion Diagnostic and Therapeutic Product Approval Approaches (FDA, 2014)

Scenario	Key Requirements in Approval Process
Novel therapeutic product	IVD companion diagnostic development and approval should be conducted simultaneously with the therapeutic product. Use of the IVD companion diagnostic should be included in the labeling of the therapeutic product. The IVD companion diagnostic must be validated and meet the applicable standard for safety and effectiveness for the use indicated in the therapeutic product labeling. Where the IVD companion diagnostic is deemed essential to the safe and effective use of the novel therapeutic, approval of the therapeutic product will not be granted until the IVD companion diagnostic is approved/cleared for that indication.
Approval of a therapeutic product without an approved/cleared IVD companion diagnostic	For new therapeutic products to treat serious or life-threatening conditions: When the benefits of using the therapeutic product outweigh the risks from the lack of an approved IVD companion diagnostic, FDA can decide to approve the therapeutic product prior to approval of an IVD companion diagnostic. The decision is determined by FDA during product review. For previously approved therapeutic products: Supplements to an approved therapeutic product labeling are not approved until the IVD companion diagnostic is approved/cleared, with the exception of revisions to the labeling to address a serious safety issue.

Table 6.3 IVDR Risk Classes

Class	Risk level		Examples
A	Patient–Low	Public – Low	Clinical chemistry analyser
B	Patient-Moderate	Public – Low	Urine test strips
C	Patient–High	Public – Moderate	Companion diagnostics
D	Patient–High	Public - High	HIV-testing

6.6.2 EUROPEAN PERSPECTIVE

Currently in Europe, the IVDD has a list-based classification system, which was written prior to the development of companion diagnostics. This has resulted in companion diagnostics falling into the lowest risk class (general IVDs). Thus, the companion diagnostics, as self-declared IVDs, do not require NB involvement.

The IVDR has acknowledged this gap by including the following definition for companion diagnostic:

"'companion diagnostic' means a device which is essential for the safe and effective use of a corresponding medicinal product to:

- *identify, before and/or during treatment, patients who are most likely to benefit from the corresponding medicinal product or*
- *identify, before and/or during treatment, patients likely to be at increased risk of serious adverse reactions as a result of treatment with the corresponding medicinal product."*

The IVDR also defines the classification of most companion diagnostic as class C, which requires conformity assessment by an NB. Table 6.3 provides a summary of the risk-based IVD classes as detailed in the IVDR.

Under the IVDR, the conformity assessment of a class C IVD shall require assessment of the quality management system and the sampling of the technical documentation by a NB. For companion diagnostics, the involvement of a consultation with the EMA or an EU medicines competent authority shall also be required. The exact details of this process will be provided in the implementing acts to be published by the EU Commission before 2022.

6.7 CONCLUSIONS

The development of combination products continues to show strong growth, with an estimated global value of 115.1 billion USD in 2019 (BBC Research, 2015). From a global perspective, the regulatory approval process for combination products is complex. However, much work has been done in recent years to provide guidance and ensure better definition. The US regulatory requirements for combination products, including companion diagnostics, are the most advanced in the world. The FDA office of combination products provides significant support to manufacturers, coordinates activities among FDA centers, and is a role model for other territories. In the EU, the process for market access is undergoing the biggest change since the implementation of the original directives. The upcoming MDR and IVDR will provide guidance for classification and regulation of combination products, closing the current regulatory gap in Europe. Other regulatory authorities around the world are further behind, with most lacking a regulatory framework for this new class of medical products.

With many anticipated changes in the future, the only certainty is that this area of health technology remains an exciting and challenging sector for medical device manufacturers, pharmaceutical companies, and regulators across the globe. To ensure mankind benefits from these healthcare advancements, these increasingly changing and novel technologies must be understood and regulated in a timely manner.

REFERENCES

BBC Research. (2015). Global markets for drug–device combinations. https://www.bccresearch.com/market-research/pharmaceuticals/drug-device-combinations-markets-report-phm045d.html

FDA. (2014). Guidance for industry and FDA staff: In vitro companion diagnostic devices. https://www.fda.gov/downloads/medicaldevices/deviceregulationandguidance/guidancedocuments/ucm262327.pdf

FDA. (2017a). CFR—code of Federal Regulations Title 21. https://www.accessdata.fda.gov/scripts/cdrh/cfdocs/cfcfr/CFRSearch.cfm?fr(3.2

FDA. (2017b). Medical devices. http://www.fda.gov/MedicalDevices/default.htm

MHRA. (2008). Bulletin no. 4 conformity assessment procedures (medical devices regulations). http://webarchive.nationalarchives.gov.uk/20130513202440/http://www.mhra.gov.uk/home/groups/es-era/documents/publication/con007492.pdf

The Council of the European Communities. (1990). Directive 90/385/EEC of 20 June 1990 on the approximation of the laws of the Member States relating to active implantable medical devices. http://eur-lex.europa.eu/LexUriServ/LexUriServ.do?uri(CONSLEG:1990L0385:20071011:en:PDF

The Council of the European Communities. (1993). Directive 93/42/EEC of 14 June 1993 concerning medical devices. http://eur-lex.europa.eu/LexUriServ/LexUriServ.do?uri(CONSLEG:1993L0042:20071011:EN:PDF

The European Commission. (2015). Manual on Borderline and Classification in the Community Regulatory Framework for Medical Devices Version 1.17. http://ec.europa.eu/geninfo/query/index.do?queryText=Manual+on+borderline+and+classification+in+the+community+regulatory+framework+for+medical+devices&summary=summary&more_options_source=global&more_options_date=*&more_options_date_from=&more_options_date_to=&more_options_language=en&more_options_f_formats=*&swlang=en

The European Commission. (2017). Nando (new approach notified and designated organisations) information system. http://ec. europa. eu/growth/tools-databases/nando/

The European Parliament and the Council of the European Union. (1998). Directive 98/79/EC of 27 October 1998 on in vitro diagnostic medical devices. http://eur-lex.europa.eu/legal-content/EN/TXT/PDF/?uri(CELEX:31998L0079&from(EN

The European Parliament and the Council of the European Union. (2001). Directive 2001/83/EC of 6 November 2001 on the Community code relating to medicinal products for human use. https://ec.europa.eu/health//sites/health/files/files/eudralex/vol-1/dir_2001_83_consol_2012/dir_2001_83_cons_2012_en.pdf

The European Parliament and the Council of the European Union. (2004). Regulation (EC) no. 726/2004 of 31 March 2004 laying down Community procedures for the authorisation and supervision of medicinal products for human and veterinary use and establishing a European Medicines Agency. http://eur-lex.europa.eu/LexUriServ/LexUriServ.do?uri(OJ:L:2004:136:0001:0033:en:PDF

The European Parliament and the Council of the European Union. (2007). Regulation (EC) no. 1394/2007 of 13 November 2007 on advanced therapy medicinal products and amending Directive 2001/83/EC and Regulation (EC) No. 726/2004. https://ec.europa.eu/health//sites/health/files/files/eudralex/vol-1/reg_2007_1394/reg_2007_1394_en.pdf

The European Parliament and the Council of the European Union. (2017a). Regulation (EU) 2017/745 of 5 April 2017 on medical devices, amending Directive 2001/83/EC, Regulation (EC) No. 178/2002 and Regulation (EC) No. 1223/2009 and repealing Council Directives 90/385/EEC and 93/42/EEC. http://eur-lex.europa.eu/legal-content/EN/TXT/PDF/?uri(OJ:L:2017:117:FULL&from(EN

The European Parliament and the Council of the European Union. (2017b). Regulation (EU) 2017/746 of 5 April 2017 on in vitro diagnostic medical devices and repealing Directive 98/79/EC and Commission Decision 2010/227/EU. http://eur-lex.europa.eu/legal-content/EN/TXT/PDF/?uri(OJ:L:2017:117:FULL&from(EN

CHAPTER

Food

7

Roger Clemens*, Christy Kadharmestan**

**University of Southern California School of Pharmacy, International Center for Regulatory Science, Los Angeles, CA, United States;*

***College of Law, Michigan State University, East Lansing, MI, United States*

"I aimed at the public's heart, and by accident I hit it in the stomach."

– Upton Sinclair

An Overview of FDA Regulated Products. http://dx.doi.org/10.1016/B978-0-12-811155-0.00007-7

CHAPTER OBJECTIVES

After reading this chapter, the reader will be able to:

- identify the primary federal regulatory agencies responsible for food safety,
- describe the GRAS process,
- recognize emerging regulatory challenges in GMO safety assessment,
- identify differences between assessing the safety of food additives, color additives, and flavoring agents, and
- describe the complexities and challenges associated with updating nutrition labeling.

7.1 FOOD REGULATORY HISTORY

The United States provides one of the safest food supplies in the world. Many regulatory agencies, food industry entities, private organizations, and consumers aggressively monitor a spectrum of factors included in the food safety continuum. Perhaps, one of the most influential public outcries for food safety regulations was spurred by Upton Sinclair's "The Jungle" (Sinclair, 1906). This 1906 novel depicted a scandalous, harsh, unsanitary meatpacking operation in Chicago. Congressional action followed, and with the executive pen of President Roosevelt, food safety was addressed in two key legislative pieces—the Pure Food and Drug Act, and the Meat Inspection Act. These acts introduced the concepts of food adulteration and misbranding, which remain central to contemporary regulations that impact the food supply.

There is a spectrum of factors that contribute to food contamination. These factors include: (1) accidental contamination, (2) economically motivated adulteration, (3) intentional contamination, (4) counterfeit/diversion tampering, and (5) disgruntled employees/sabotage.

In the interim of more than 100 years, a myriad of amendments to the 1938 Food Drug and Cosmetic Act and other federal acts were enacted by Congress, typically in response to emerging food and drug safety issues, including manufacturing practices, marketing tactics, and advertising claims. Some of these reactive regulations include the Food Additives Amendment and the Delany proviso in 1958, the Color Additive Amendment (1960), the Toxic Substances Control Act (1976), the Infant Formula Act (1980) and its amendments (1986), the Safe Drinking Water and Toxic Enforcement Act (1986; also known as Proposition 65 in California), the Organic Food Production Act (1990) and the National Labeling and Education Act that same year, the Dietary Supplements and Health Education Act (1994), and the Public Health and Security Preparedness and Response Act (2002).

The Infant Formula Act was initiated in response to an inappropriate infant formula with a sodium-reduced formulation that markedly decreased chloride in the Neo-Mull-Soy products manufactured by Syntex in the late 1970s. Consumption of this product by infants resulted in hypochloremia and hypokalemia as reported by the Centers for Disease Control in 1978–79. Fortunately, chloride supplementation led to the recovery of those infants affected. Until this time, the Food and Drug Administration (FDA) had not specified the level of chloride as recommended by the American Academy of Pediatrics. While the FDA required specific product labeling of food intended for infant use in 1941, the agency's guidelines for infant formula (1962–66) only specified seven vitamins and four minerals. Thus, in 1980, the Infant Formula Act included a "must" composition of infant formula intended for healthy, term infants, and with its amendments established enforceable quality control factors and good manufacturing practices (GMPs) (21 CFR § 106), recall procedures, and the requirement for premarketing clinical trials and notification process. Today, the "must" composition list includes 30 nutrients, minimum levels for all of these nutrients, and maximum levels for 10 of these (protein, fat, vitamin A, vitamin D, iron, iodine, selenium, sodium, potassium, chloride) per 21 CFR § 107.

One of the more contentious and controversial regulations is California's Proposition 65 that was approved by the voters of California in a 1986 general election. This state law required the governor, then George Deukmejian, to create and maintain "a list of those chemicals known to the state to cause cancer or reproductive toxicity" (California, 2018). The current list (as of December 31, 2017) identifies 973 substances that presumably represent a health hazard such as acrylamide, which is found in baked foods and roasted coffee. It is important to note that this substance, a product of the amino acid asparagine and some types of carbohydrates when exposed to elevated temperatures, has not been shown to be harmful in humans. The regulatory action taken in California reflects limited data from rodent models as studied by the International Agency on Cancer Research and the National Toxicology Program that classify acrylamide as a potential or possible carcinogen. Interestingly, the U.S. Environmental Protection Agency (EPA), which regulates acrylamide in water, established an acceptable level of not to exceed 1 ppm. The list includes FDA-approved medications, FDA-approved food additives, compounds that occur naturally in foods, beverages and cosmetics, and pesticides, regardless of their origin (natural or synthetic). Caffeic acid, a natural component of plants, such as coffee, cinnamon, anise, thyme, sage, sunflower seeds, apricots, prunes, barley, and rye, has shown some potential health benefits as evidenced by limited in vitro and in vivo studies (Sato, Itagaki, & Kurokawa, 2011). However, it is in the 1994 listing based on evidence presented by an authoritative body, for example, the National Institute for Occupational Safety and Health (NIOSH), the National Toxicology Program (NTP), or International Agency for Research on Cancer (IARC).

The first proactive food safety legislation, the Food Safety Modernization Act (FSMA), was enacted in 2011. This Act increased the Congressional authority and accountability of the FDA in ways that affect both domestic and international food producers. Fig. 7.1 depicts the reactive and dynamic regulatory changes that impacted food safety and quality since 1906.

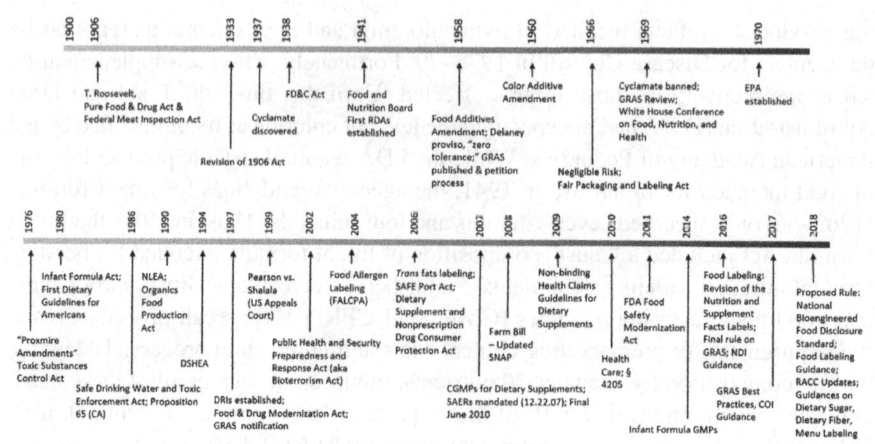

FIGURE 7.1 Food law and nutrition guidance milestones.

7.2 WHICH AGENCY IS RESPONSIBLE FOR REGULATING FOOD?

In the United States, several federal agencies have shared responsibilities, jurisdiction, and authority with respect to food supply and its safety.

7.2.1 FDA

The FDA, specifically the Center for Food Safety and Applied Nutrition (CFSAN), regulates foods, dietary supplements, and cosmetics, along with its other areas of responsibility. It is important to note that food, as defined in the Food, Drug, and Cosmetic Act (FDCA) § 201, refers to:

(1) articles used for food or drink for man or other animals,
(2) chewing gum, and
(3) articles used for components of any other such article.

In addition, the FDA regulates dietary supplements, both the finished products and ingredients. The agency's primary responsibilities relative to dietary supplements are outlined in the 1994 Dietary Supplements and Health Education Act (Dietary Supplements, 2018).

In December 2015, the FDA created the Office of Dietary Supplement Programs (ODSP), elevating the program from its previous status as a division under the Office of Nutrition Labeling and Dietary Supplements. The latter division is now known as the Office of Nutrition and Food Labeling. The initial objectives of the ODSP are to assure the quality of dietary supplements are produced under GMPs, to address marketing practices and composition issues, such as unapproved ingredients and pharmaceutical agents that compromise consumer health, and to assure the dietary supplement industry provides credible scientific evidence for health claims that point to structure–function of the human body, while steering clear of making direct or implied drug claims.

7.2.2 USDA

With respect to food safety, the United States Department of Agriculture (USDA) assures the commercial supply of meat, poultry, and egg products are safe, wholesome, and properly labeled and packaged. Unlike foods under FDA jurisdiction, products and their respective labels under the USDA authority require premarket review and approval through the Food Safety and Inspection Service (FSIS).

7.2.3 FSIS

The Food Safety and Inspection Service (FSIS) is the public health service under the USDA. One of many important areas of FSIS responsibilities is its labeling guidance, policies, and inspection methods that will ensure that all labels are truthful and not misleading. The four major components of the FSIS labeling program are presented in 9 CFR § 412.1(c). Importantly, the labeling of foods that contain more than 3% meat or 2% poultry products, which are under FSIS jurisdiction, must be approved prior to commercial distribution within the United States (21 U.S.C. 607(d) and 21 U.S.C. 457(c)).

7.2.4 EPA

The Environmental Protection Agency (EPA) was granted authority under the Food Quality Protection Act of 1996 and has several responsibilities with respect to food safety, water quality, and pesticide application. The EPA conducts comprehensive reviews of pesticides applied to agricultural products and detection in foods and establishes tolerances for pesticide residues. The maximum limits of pesticides in foods are based on classic toxicology of these contaminants, both man-made and nature-made, frequency of application and potential consumer exposure, and a practical margin of safety to minimize adverse events that may be associated with the pesticide contact and consumption.

7.2.5 FWS

The Fish and Wildlife Service (FWS), under the Department of the Interior, oversees the safety of edible animals and plants in the outdoors as well as a spectrum of flora and fauna in the environment. For example, FWS is responsible for fish and wildlife conservation, the national fish hatchery system, the products of which are typically consumed by people and animals.

7.2.6 TTB

The Alcohol and Tobacco Tax and Trade Bureau, better known as the Tax and Trade Bureau (TTB), was a component of the Bureau of Alcohol, Tobacco and Firearms (ATF). The Bureau was split into two operations, the TTB and the ATF, in 2003 under the 2002 Homeland Security Act. According to a Memorandum of Agreement (MOA) with the FDA, the TTB Beverage Alcohol Laboratory (BAL) is responsible for analyzing alcohol beverage products for limited and prohibited compounds. The laboratory enforces these restrictions for alcohol beverages as per 21 CFR § 189: Substances Prohibited from Use in Human Food. BAL also assesses the presence of allergenic fining agents that may be used in wine prior to bottling. Especially important for consumers allergic to egg and milk products, BAL initiated a program to evaluate specific food allergens in malt beverages, wines, and distilled spirits.

7.2.7 DEA

The Drug Enforcement Administration (DEA) enforces the controlled substances laws and regulations of the United States. With respect to foods, the DEA, in collaboration with the FDA, evaluates foods and food ingredients for possible contamination with illicit agents or pharmaceutical substances. Several recent investigations focused on synthetic cannabinoids in baked goods and state laws that permit marijuana in foods. For example, in 2015, the state of Colorado permitted the claim of 10 mg THC (tetrahydrocannabinol) in edible confections as indicated by a universal symbol. Below the symbol, the statement "Contains Marijuana. For Medical Use Only. Keep out of the reach of children." From a DEA perspective, if these commercial products do not cross state lines, then Federal law that prohibits recreational and medical cannabis may not be violated under the Controlled Substances Act despite the prosecution mandate (Controlled Substances Act, 21 U.S.C. § 811). However, on January 4, 2018, the U.S. Department of Justice issued a memo that reinforced the Act, which prohibits the cultivation, distribution and possession of marijuana (21 U.S.C. § 801 *et seq*)

7.3 KEY FOOD REGULATORY STANDARDS

It is important to note that the FDA specifies standards for different product categories. For example, a dietary supplement refers to a product "intended to supplement the diet" that contains one or more of the following: (a) a vitamin, mineral, (b) herb or botanical, (c) amino acid (FD&C Act §201(ff)(1)). Medical food is formulated to be consumed or administered enterally under the supervision of a physician and is intended for the dietary management of a specific disease or condition (21 CFR §101.9(j)(8)). Examples include inborn or genetic errors of amino acid metabolism, such as branched-chain amino acids isoleucine, leucine, valine (maple syrup urine disease), and phenylalanine (phenylketonuria).

A color additive is any dye, pigment, or substance that can impart color when added or applied to a food (FD&C Act § 201(t)). Within the United States, colors are classified as certified colors, or color exempt from certification. From the consumer's perspective, certified colors are considered synthetic, and exempt colors are considered natural. Regardless of color classification, each color intended for applications in foods must meet the standards established in the 1960 Color Additives Amendment to the FD&C Act. Food colors are not food additives and must be considered safe for their intended uses and approved by the FDA prior to commercial application. Importantly, the Color Additive Amendment includes the Delaney clause, requiring safety data, which indicate that the substance does not cause cancer in animals or humans. In addition, the application of food colors must not promote potential "deception of the consumer." See 21 U.S.C. § 379e(b)(6) for additional information.

According to the Food Additive Amendment of 1958, food additives include substances that are added to food for specific purpose. For example, processing aids (21 CFR §101.100(a)(3)) are substances that are added to a food for their technical or functional effect in the processing but are present in the finished food at insignificant levels. These are substances, such as organic acids and chlorine washes, which are used to clean fresh fruit and vegetables, and rennet, which is used in cheese production. The USDA has approved many processing aids for use in meat and poultry, which may control pH (e.g. ammonium hydroxide) and bacteria in chill water (chlorine gas). None of the processing aids approved by the FDA and USDA is required to be stated in the ingredient declaration. This is contrary to food labeling requirements intended for companion animals and livestock feeds.

Standard of identity (SOI), established by the FDA, refers to mandatory requirements associated with common or usual names of approximately 300 foods. For example, 21CFR169.140 contains Standard of Identity for mayonnaise. These requirements promote honesty, fair dealing, and product consistency for the benefit of consumers. See 21 CFR § 130–169 for a complete list of foods with a standard of identity: Primary Food Regulation.

As discussed in Chapter 1, the US Code of Federal Regulations (CFR) encompasses 50 different titles or areas of responsibility. Relative to food regulations, Titles 7, 9, and 21 are three key resources.

Title 7 focuses on agriculture (USDA), which entails Child Nutrition Programs. These programs include National School Lunch Program, School Breakfast Program, Child and Adult Care Food Program, Summer Food Service Program, Fresh Fruit and Vegetable Program, and Special Milk Program, which are directed to deliver the best nutrition and provide nutrition education to millions of school-age children.

Title 9 addresses animal health and animal products. For example, within this title, livestock feed and health issues, such as chronic wasting disease, avian flu, scabies in cattle, sanitation control, and import/export regulations are stipulated.

Title 21 contains nearly 1500 parts. Within this title, there are parts dedicated to human food regulations, such as imports and exports, food facility registration, enforcement policies, color additives, food labeling, food standards, direct and indirect food additives, and many other aspects that affect the safety and quality of foods designated for human consumption.

Named after Representative James Delaney (D-NY; 1901–87), the Delaney Clause prohibits the approval of any food additive found to induce cancer in man or animal or found to induce cancer in appropriate tests. This clause appears in the Food Additives Amendment of 1958, the Color Additives Amendment of 1960, and the Animal Drug Amendments of 1968. This clause does not include toxins, carcinogens or mutagens that are innate to foods, such as those that dominate plant-based foods (e.g., fruits, vegetables).

Because today's analytical methods and animal testing protocols are far more sensitive than those available or even considered in the 1950s and 1960s when the Delaney Clause came into effect, it became unreasonable to expect zero residues of the culprit additives. Hence, the 1996 Food Quality Protection Act eliminated the zero tolerance for carcinogenic pesticide residues to a "reasonable certainty that no harm will result" based on a cumulative risk assessment from a multiple exposures approach (21 U.S.C. § 346a(b)(2)(A)(ii)).

7.3.1 FOOD SAFETY MODERNIZATION ACT

The FSMA of 2011 marked the first proactive legislation intended to improve the safety of the domestic food supply and imported foods. The major five elements of this legislation include (1) preventive controls, (2) inspection and compliance, (3) imported food safety, (4) response, and (5) enhanced partnerships. Preventive controls include mandates that aim to reduce the likelihood of foodborne adverse events. The inspection and compliance element includes adopting innovative approaches to inspections based on potential risks. Imported food safety measures incorporate registration and inspection of foreign food suppliers and their respective facilities to assure compliance with food safety standards. The response element provides the FDA with a new authority to mandate food recalls and close a production facility, if warranted. Expanded partnerships encourage collaboration among all federal, state, and local agencies to improve and provide consistency in food safety personnel training (FDA, 2016).

7.3.2 COTTAGE FOOD OPERATIONS (STATE-SPECIFIC)

Cottage laws are intended to allow individuals to commercialize private home-prepared foods, which are presumably nonhazardous. Effective January 1, 2013, the State of California specified "cottage food operations" to meet specific criteria, such as requirements consistent with the State's Health and Safety Code related to foods on the approved list. The operators must complete a food processor course in 3 months following facility registration, implementation of sanitary operations, establish state and federal compliant food labels, and operate within established gross annual sales limits (<$50,000 after 2015). The approved food list includes 31 food products, such as confections, fruit pies, herb blends, and dehydrated vegetables that, if appropriately managed, are unlikely to precipitate a health hazard (Cottage Food Operations, 2016).

7.3.3 FOOD DEFECT ACTION LEVELS (21CFR§110.110)

Food Defect Action Levels refer to levels of natural or unavoidable defects in foods that present no health hazards in humans. The FDA acknowledges these action levels because it is economically impractical to grow, harvest, or process raw products that are totally free of nonhazardous, naturally occurring, unavoidable defects. Examples of commodities and their limits of natural defects include, but are not limited to, apple butter (four or more rodent hairs per 100 g product), chocolate (more than 60 insect fragments per 100 g product), green coffee beans (more than 10% of insect-infested or insect-damaged beans), and peanut butter (average 30 or more insect fragments per 100 g product). A complete listing of foods and their defect action levels may be found at http://www.bodek.com/fda_action_levels_p.pdf that was last updated in 2009.

7.3.4 FLAVOR SAFETY

The safety of flavor additives has been reviewed as generally recognized as safe (GRAS) ingredients by the Flavor Extract Manufacturers Association (FEMA) under a memorandum of understanding (MoU) with the FDA since 1959. This group of experts leverages common approaches in safety evaluations, including the use of metabolic studies and structural relationships, while considering duration and frequency of oral exposure. The biennial report on flavor additives is published in *Food Technology*, a publication sponsored by the Institute of Food Technologists, is also available at the FEMA web site (http://www.femaflavor.org/fema-gras).

7.4 ADULTERATION

There are regulatory criteria by which foods are considered adulterated (Food Drug and Consmetic Act, 2018). Fundamentally, a food would be deemed adulterated if it contains any poisonous or deleterious substance which may be detrimental to human health. This standard does not apply to naturally occurring substances, which of course are not intentionally added. A product may also be considered adulterated if it contains a pesticide, food additive, color additive, or animal drug that may be considered unsafe under Section 406 of the FDCA. This section established tolerances of poisonous substances in foods (21 CFR § 109.4).

7.5 MISBRANDED

Misbranded foods are those with false or misleading labeling according to Section 403 of the FDCA. This section also defines misbranding to include those foods that are imitations of other foods unless identified as "imitation," or purported to represent products that have a standard of identity, yet do not meet the regulated criteria.

7.6 WHAT IS HEALTHY?

In 1994, the FDA established criteria for the use of the word "healthy" relative to food product claims. Those criteria, fundamentally based on the existing dietary guidelines, included specific labeling claims, such as total fat, saturated fat, sodium, cholesterol, and potential beneficial nutrients such as vitamin C or calcium in servings customarily consumed (1994; 21 CFR § 101.65(d)). The FDA is reconsidering these labeling guidelines, since the science no longer considers total dietary fat and cholesterol critical factors for health. For example, the preponderance of scientific evidence suggests diets that include avocados and nuts may reduce the risk of cardiovascular disease. As clinical and epidemiological evidence prompted changes in dietary recommendations noted in 2015–20 Dietary Guidelines for Americans, the FDA is working to redefine the term "healthy" (Food and Drug Administration, 2018a, b).

7.7 NATURAL

Another controversial topic in food labeling and product claims is use of the word "natural." In 1993, the FDA published a policy indicating that "natural" refers to substances that are not synthetic or artificial. The use of this term is permitted by the FDA if the food did not contain added color, artificial flavors, or synthetic substances. This policy differs from the 1982 policy by the USDA stating that "natural" could be applied if the food did not contain any artificial flavor, coloring ingredient, or chemical preservative, or was only minimally processed. As with the term "healthy," the FDA is working to clearly define the term "natural."

In the United States recently, consumers have been calling for replacement of synthetic (certified) colors with natural (exempt) colors (21 CFR §73) for the fear of adverse events, particularly among children. Although frequently cited, the results from studies on potential health issues, such as hyperactivity and ADHD, among susceptible children have not been corroborated (Bateman, Warner, & Hutchinson, 2004; McCann, Barrett, & Cooper, 2007). In 2011, the FDA convened the Food Advisory Committee to address potential health issues that may be associated with artificial food colors. Upon an extensive review of more than 30 studies on food dyes, and a critical assessment of biological and environmental factors that may contribute to ADHD, the FDA concluded that additional systematic studies are required to better understand interactions with nutrients, medications, and among/between artificial colors. The situation is especially puzzling in that, while the use of these colors continues to decline, the percent of children diagnosed with ADHD continues to increase (Arnold, Lofthouse, & Hurt, 2012; Centers for Disease Control, 2018).

With respect to foods directed at companion animals and livestock, the term "natural" is clearly defined. Under the authority of the Center for Veterinary Medicine (CVM), the Association of American Feed Control Officials (AAFCO) codified "natural" as "a feed or feed ingredient derived solely from plant, animal or mined sources" (AAFCO, 2018).

7.8 ORGANIC

Under the USDA National Organic Program and stipulated in the Act, "organic" is a labeling term that indicates that the food or other agricultural product has been produced through approved methods (National Organic Program, 2018). Under this Act, all agricultural products labeled "organic" must be derived from farms or facilities that are certified by an appropriate Agency approved and accredited by the USDA. It does not state or imply improved safety or nutritional quality over like-foods from conventional agricultural practices (Environmental Protection Agency, 2018).

There are three basic categories of organic foods. 100% Organic may be declared if the product contains 100% organic ingredients. The descriptor "organic" may be stated on the label if the product contains a minimum of 95 percent organic ingredients. Finally, the phrase "Made with Organic ______" can be used to label a product that contains at least 70% organically produced ingredients. See 7 U.S.C. Chapter 94: Organic Certification for additional information.

7.9 NUTRITION LABELING (USA)

The National Labeling and Education Act of 1990 (NLEA; Public Law 101-535) provided the FDA with specific authority to require nutrition labeling of most foods. Specific components and format of the ensuing food labels were codified in Title 21: 21 CFR § 101.2 (label review), 21 CFR § 101.9(d) (general format and print size), 21 CFR § 101.9(b)(1) (serving size), 21 CFR § 101.9(b)(8) (servings per container), 21 CFR 101.9(c) (amount per serving), 21 CFR § 101.9(c)(1) (calories), and 21 CFR § 101.9(c)(7) & (8) & (9) and (d)(6) & (7) (percent daily value).

Following publication of the 2015–20 Dietary Guidelines for Americans, the FDA published new nutrition facts panel rules to reflect those guidelines (FDA Nutrition Panel Changes, 2018). Features of the final nutrition facts label (shown in Fig. 7.2) include a refreshed design and updates to improve information access for consumers. For example, the font size for "calories," "servings per container," and the "serving size" were changed, and the calories declaration is emboldened. Another enhancement is the declaration of the actual amounts and Daily Value of vitamin D, calcium, iron and potassium. These were considered nutrients of concern in the 2015–20 Dietary Guidelines for Americans (DGA). One other change was the footnote which included the meaning the Daily Value, and its relative value to a daily diet of 2,000 Calories.

A comparison of old and new nutrition facts panels is shown in Fig. 7.3.

As our understanding of nutrition changes, so does this new labeling format and its content (Food Labeling Revision, 2016). In particular, "Added Sugars" was defined, and a daily value established. This information must be declared in grams as well as percent daily value (DV). The DV should not exceed 10% of total daily energy intake.

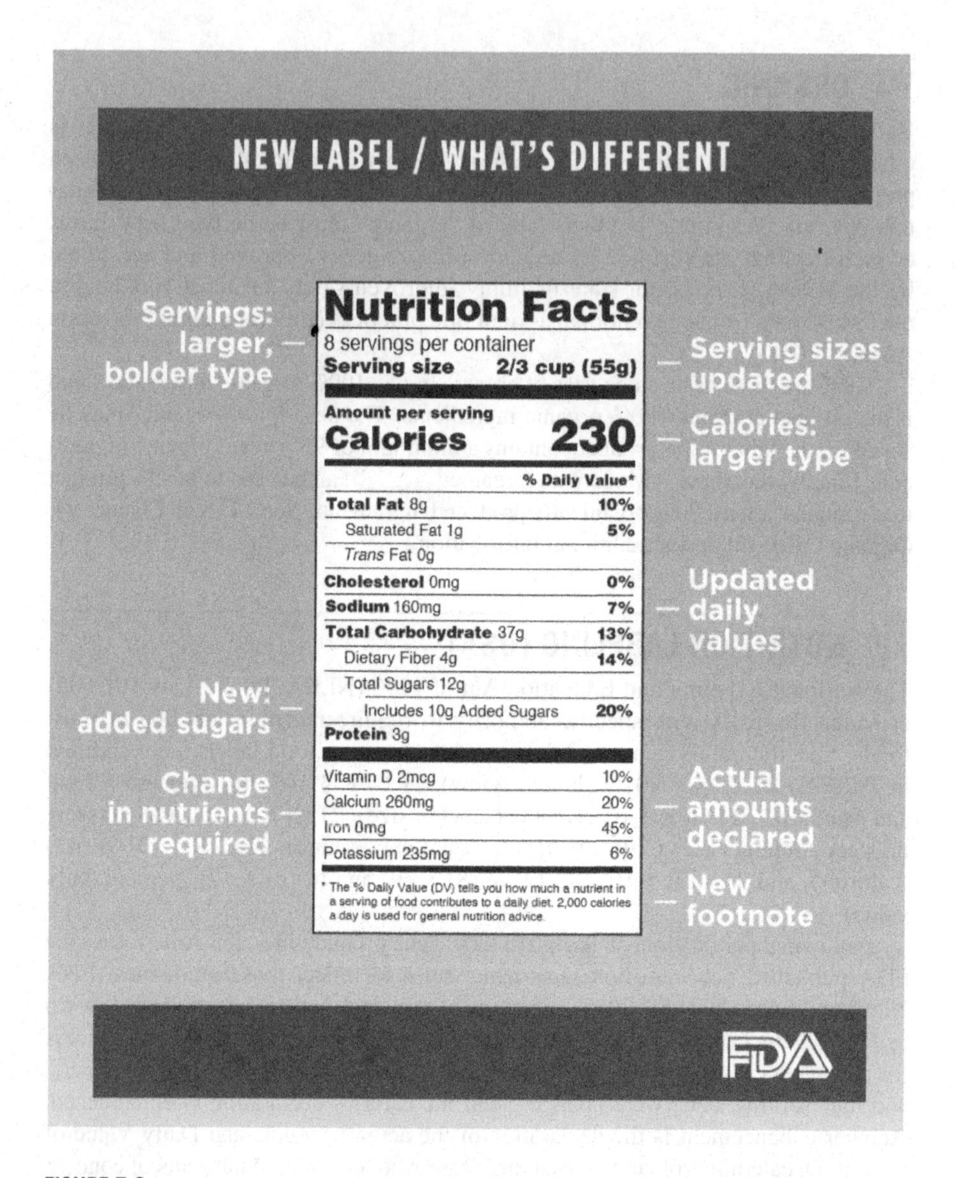

FIGURE 7.2

Highlights of what's different on the new label (https://www.fda.gov/Food/GuidanceRegulation/GuidanceDocumentsRegulatoryInformation/LabelingNutrition/ucm385663.htm).

Since the DGA did not establish an upper limit of total dietary fat, "calories from fat" was removed from the panel. Daily values for nutrients, such as sodium, dietary fiber, and vitamin D, were updated based on findings presented in the DGA.

Nutrition Facts

Serving Size 2/3 cup (55g)
Servings Per Container About 8

Amount Per Serving	
Calories 230	Calories from Fat 72
	% Daily Value*
Total Fat 8g	**12%**
Saturated Fat 1g	**5%**
Trans Fat 0g	
Cholesterol 0mg	**0%**
Sodium 160mg	**7%**
Total Carbohydrate 37g	**12%**
Dietary Fiber 4g	**16%**
Sugars 1g	
Protein 3g	
Vitamin A	10%
Vitamin C	8%
Calcium	20%
Iron	45%

* Percent Daily Values are based on a 2,000 calorie diet. Your daily value may be higher or lower depending on your calorie needs.

	Calories:	2,000	2,500
Total Fat	Less than	65g	80g
Sat Fat	Less than	20g	25g
Cholesterol	Less than	300mg	300mg
Sodium	Less than	2,400mg	2,400mg
Total Carbohydrate		300g	375g
Dietary Fiber		25g	30g

Nutrition Facts

8 servings per container
Serving size 2/3 cup (55g)

Amount per serving	
Calories	**230**
	% Daily Value*
Total Fat 8g	**10%**
Saturated Fat 1g	**5%**
Trans Fat 0g	
Cholesterol 0mg	**0%**
Sodium 160mg	**7%**
Total Carbohydrate 37g	**13%**
Dietary Fiber 4g	**14%**
Total Sugars 12g	
Includes 10g Added Sugars	**20%**
Protein 3g	
Vitamin D 2mcg	10%
Calcium 260mg	20%
Iron 8mg	45%
Potassium 235mg	6%

* The % Daily Value (DV) tells you how much a nutrient in a serving of food contributes to a daily diet. 2,000 calories a day is used for general nutrition advice.

FIGURE 7.3

Old-new label format comparison (https://www.fda.gov/Food/GuidanceRegulation/GuidanceDocumentsRegulatoryInformation/LabelingNutrition/ucm385663.htm).

Another significant change that is reflected in nutrient declaration is food serving size (see Fig. 7.3). The updated labeling is intended to reflect more realistic amounts of foods and beverage that are typically consumed. Data from Nationwide Food Consumption Surveys and National Health and Nutrition Examination Surveys (NHANES) indicated consumption patterns in the United States have changed; thus, the FDA contended it was important to amend the Reference Amounts Customarily Consumed (RACCs) relative to 1993 data (Food Labeling Serving, 2016a, b). For example, a bottled beverage becomes a single serving regardless of the size or volume. The RACC of yogurt was reduced to 6 oz. (170 g) from 8 oz.

The final ruling indicated the FDA intended to exercise enforcement discretion either July 26, 2018 or July 26, 2019. In May 2018, the FDA extended these dates.

Small manufacturers (<$10MM annual sales) must comply before January 1, 2021. Large manufacturers (>$10MM annual sales) must comply by January 1, 2020.

7.10 FUNCTIONAL FOODS (USA)

The functional foods concept was initiated in the United States by Stephen DeFelice and Steve McNamara in 1989. Actually, they suggested the term "nutraceuticals" based on the 1983 Orphan Drug Act that was established to permit the distribution of drugs destined to treat rare medical conditions that affect a small number of people (<200,000) in the United States. Nutraceuticals are substances that are a food or part of a food that provides medical and/or health benefits, including the prevention and treatment (these are "drug" terms) of disease (Roberfroid, 2000). As of this writing, the United States has yet to promulgate functional foods standards.

There does not appear to be any statutory harmonized definition of functional foods throughout Europe, yet Japan established Foods for Special Dietary Health Use (FOSHU), which requires government review of scientific evidence and approve such foods as a component of the traditional dietary pattern. An excellent review on functional foods was published by the Institute of Food Technologists in 2005 (Clydesdale et al., 2005). A follow-up review points out that perhaps the definition of functional foods should be changed and advocates those changes in support of purported bioactives (Martirosyan and Singh, 2015). With scientific ground work advancing to define DRIs for bioactives, FDA may establish a functional foods category and add legitimacy to the marketing of such food products (Lupton, Atkinson, & Chang, 2014;Ranard, Jeon, & Mohn, 2017).

A more recent report outlined several contemporary challenges and opportunities in the functional foods arena, including the impact of processing on the functional properties of functional foods (Rodgers, 2016).

7.11 HEALTH CLAIMS

Fundamentally, health claims describe a relationship between a food or food ingredient and a health outcome or reduced risk of developing an adverse health condition. In the United States, there are several categories of claims, namely health claims authorized under the National Labeling and Education Act (NLEA, 1990), and health claims based on authoritative statements under the Food and Drug Administration Modernization Act (FDAMA, 1997). This Act included nutrient content claims and structure/function claims.

Nearly a decade later in the Pearson v Shalala case (Roberfroid, 2000), the United States Court of Appeals, District of Columbia Circuit, the Court rejected the FDA's argument to deny requested health claims for dietary supplements and rendered the position that producers of these products could make health claims based on significant scientific agreement in four areas. Those areas included (1) consumption of antioxidant vitamins may reduce the risk of certain kinds of cancers, (2) consumption of fiber may reduce the risk of colorectal cancer, (3) consumption of omega-3 fatty acids may reduce the risk of coronary heart disease, and (4) 0.8 mg of folic acid in a dietary supplement is more effective in reducing the risk of neural tube defects than a lower amount in foods in common form (US, 1999).

Under NLEA, the FDA could authorize health claims following an exhaustive review and evaluation of the scientific evidence initiated by the Agency or those petitioning the claim. On the other hand, under FDAMA of 1997, an authoritative statement by the National Academy of Sciences or other scientific body within the US government may be used to substantiate a health claim. Such authoritative statements reflect a relationship between a nutrient and a disease or health-related condition, published by a scientific body, is currently in effect, and shall indicate the agency responsible for such claim. A summary of the 12-significant scientific agreement (SSA) health claims is shown in Table 7.1.

The FDA may also authorize qualified health claims for which the interim industry guidance was issued in 2003. In these types of claims, the emerging evidence for the relationship between a food substance and risk of a disease or health-related condition may not be adequately substantiated, but permits a petition process that triggers enforcement discretion by the FDA. For enforcement discretion, the FDA notes credible evidence to support the claim, yet the Agency outlines qualifications by which the claim may be used for given foods or ingredients. A summary of more than 20 current qualified health claims is shown in Table 7.2.

In 2016, the FDA accepted a qualified health claim which states: "High-amylose maize resistant starch may reduce the risk of type 2 diabetes," although FDA has concluded that there is limited scientific evidence for this claim (Level C). The claim is only applicable to foods that contain ≥10% of the daily value (DV) for vitamins A and C, plus iron, protein and fiber, plus 10% of the DV for vitamin D or potassium per RACC Table 7.2.

As suggested in the latest qualified health claim and noted in the guidance document, there is a ranking of strength of the evidence for such claims. The qualifying language for each of three categories for such claims is shown is Table 7.3. A fourth category for qualified health claims, known as Category A, meets the same scientific requirements for an authorized health claim.

Nutrient content claims state or imply the level of a nutrient within a food product as specified under 21 CFR § 101.13. In general, such claims express the presence or absence of a nutrient which may suggest or contribute to a healthful dietary pattern, such as "low sodium," "high in oat bran," or "healthy, contains 3 grams (g) of fat."

On the other hand, there are few regulations that address structure–function claims in the United States. These kinds of claims have a foundation in the Dietary Supplement Health and Education Act (DSHEA) of 1994. As noted in the Act, there are two fundamental specifications that may describe the general well-being or may relate to a nutrient deficiency disease. For example, a structure–function claim may

Table 7.1 Significant Scientific Agreement (SSA) Health Claims[a]

	Significant Scientific Agreement Health Claims	Regulation
1.	Calcium, vitamin D, and osteoporosis	21 CFR 101.72
2.	Dietary lipids and cancer	21 CFR 101.73
3.	Dietary saturated fat and cholesterol and risk of coronary heart disease	21 CFR 101.75
4.	Dietary noncariogenic carbohydrate sweeteners and dental caries	21 CFR 101.80
5.	Fiber-containing grain products, fruits and vegetables, and cancer	21 CFR 101.76
6.	Folic acid and neural tube defects	21 CFR 101.79
7.	Fruits and vegetables and cancer	21 CFR 101.78
8.	Fruits, vegetables, and grain products that contain fiber, particularly soluble fiber, and risk of coronary heart disease	21 CFR 101.77
9.	Sodium and hypertension	21 CFR 101.74
10.	Soluble fiber from certain foods and risk of coronary heart disease	21 CFR 101.81
11.	Soy protein and risk of coronary heart disease	21 CFR 101.82
12.	Stanols/sterols and risk of coronary heart disease	21 CFR 101.83

[a] *http://www.fda.gov/Food/IngredientsPackagingLabeling/LabelingNutrition/ucm2006876.htm.*

state "calcium builds strong bones" which was similar to a tag line associated with Wonder Bread in the 1960s. Another example is "fiber maintains bowel regularity" simply describes a general well-being following the consumption of dietary fiber. However, such claims for conventional foods may also be misleading, as noted by the Federal Trade Commission (FTC) and the general public following the 2009 bird flu outbreak when Kellogg's claimed its Rice Krispies "Now Helps Support Your Childs Immunity." In 2002, the FDA issued a guidance for industry regarding structure–function claims (http://www.fda.gov/Food/GuidanceRegulation/GuidanceDocumentsRegulatoryInformation/ucm103340.htm) that specifically addresses these claims relative to dietary supplements.

7.12 BE LABELING

Few issues associated with the food supply have created as much consumer activism as genetically modified foods, also known as genetically modified foods, now termed bioengineered foods. As early as 1992, the FDA's policy on genetically modified plants applied to all foods derived from new plant varieties, including those developed through rDNA technology. Plants in the latter category were termed bioengineered foods (Statement of Policy, 1992). Nearly a decade later, the FDA published voluntary label guidance and statements regarding foods developed

Table 7.2 Health Claims: Qualified Subject to Enforcement Discretion[a]

Qualified Health Claims	Year
Atopic dermatitis	
• 100% whey-protein partially hydrolyzed infant formula and reduced risk of atopic dermatitis	2011
Cancer	
• Selenium and a reduced risk of site-specific cancers	2009
• Antioxidant vitamins C and E and reduction in the risk of site-specific cancers	2009
• Tomatoes and prostate, ovarian, gastric, and pancreatic cancers (American Longevity Petition)	2005
• Tomatoes and prostate cancer (Lycopene Health Claim Coalition Petition)	2005
• Calcium and colon/rectal cancer and calcium and colon/rectal polyps	2005
• Green tea and risk of breast cancer and prostate cancer	2011
• Selenium and certain cancers	2003
• Antioxidant vitamins and risk of certain cancers	2003
Cardiovascular disease	
• Folic acid, vitamin B6, and vitamin B12 and vascular disease	2000
• Settlement reached for health claim relating B vitamins and vascular disease	2001
Nuts and coronary heart disease	
• Walnuts and coronary heart disease	2004
• Nuts and coronary heart disease	2003
Omega-3 fatty acids	
• Corn oil and corn oil-containing products and a reduced risk of heart disease	2007
• Unsaturated fatty acids from canola oil and reduced risk of coronary heart disease	2006
• Mono unsaturated fatty acids from olive oil and coronary heart disease	2004
Cognitive function	
• Phosphatidylserine and cognitive dysfunction and dementia	2003
Diabetes	
• Psyllium husk and a reduced risk of type 2 diabetes	2014
• Whole grains and a reduced risk of diabetes mellitus type 2	2013
• Chromium picolinate and a reduced risk of insulin resistance, type 2 diabetes	2005
• High-amylose-resistant starch and a reduced risk of type 2 diabetes	2016
Hypertension	
• Calcium and hypertension, pregnancy-induced hypertension, and preeclampsia	2005
Neural tube defects	
• Folic acid and neural tube defects	2001

[a] *http://www.fda.gov/Food/IngredientsPackagingLabeling/LabelingNutrition/ucm072756.htm.*

Table 7.3 Categories for Qualified Health Claims[a]

Category	Scientific Ranking Level	Appropriate Qualifying Language[b]
B	Second	… "although there is scientific evidence supporting the claim, the evidence is not conclusive"
C	Third	"Some scientific evidence suggests … however, FDA has determined that this evidence is limited and not conclusive"
D	Fourth	"Very limited and preliminary scientific research suggests … FDA concludes that there is little scientific evidence supporting this claim"

[a] *http://www.fda.gov/Food/GuidanceRegulation/GuidanceDocumentsRegulatoryInformation/ucm053832.htm.*

[b] *The language reflects wording used in qualified health claims as to which the agency has previously exercised enforcement discretion for certain dietary supplements. During this interim period, the precise language as to which the agency considers exercising enforcement discretion may vary depending on the specific circumstances of each case.*

using bioengineering, which was updated in 2015 (Voluntary Labeling, 2015). This guidance stipulated that manufacturers who voluntarily labeled their food products would need to substantiate label statements. Earlier that year, Federal GMO labeling was proposed in the Safe and Accurate Food Labeling Act of 2015 (Safe and Accurate, 2015). This Act barred individual states from requiring labeling of GM foods and established a labeling program to certify GM-free criteria. The National Bioengineered Food Disclosure Law of 2016 stipulated the USDA, which has primary labeling authority for food items such as meat, dairy, poultry, and eggs, to develop regulations and standards to create mandatory disclosure requirements for bio-engineered foods by July 2018 (National Bioengineered, 2016). This law permitted GM disclosure using one of three options: on-package text, a USDA-created symbol, or an electronic link. The USDA posted a proposed rule for GMO labeling (843 FR 19860, 2018). The rule, if finalized, would require food manufacturers and that label foods for commerce to provide information on bioengineered (BE) foods and food ingredient content.

All foods, regardless of their course, share similar risks. Those risks include potential allergic reactions, innate or enhanced toxins and subsequent reactions, and possible increase in antinutrient effects. Fundamental principles and guidelines for safety assessment of bioengineered foods are summarized in the FDA's guidance Submissions on Bioengineered New Plant Varieties (FDA Safety, 2018).

7.13 SAFETY ASSESSMENT (GRAS)

While food additives require FDA approval before they can be marketed, substances considered to be GRAS can be used without FDA oversight.

Historically, a substance can be determined as GRAS through two pathways: Self-affirmation petition and FDA notification. The scientific procedures for the assessments are outlined in 21 CFR § 170.30(b). In general, GRAS documentation, regardless of self-affirmation or FDA notification, include key elements such as sponsor information, substance description, chemical description, production process, finished product and stability specifications, projected intended use, estimated daily intake (EDI), and a spectrum of safety data to determine acceptable daily intake (ADI). Essential safety data for GRAS panel review include classic toxicology (absorption, distribution, metabolism, excretion), toxicokinetic, genotoxicology, acute, chronic, and subchronic exposure and developmental and reproductive toxicity studies that are consistent with ICH (International Council on Harmonization) guidelines on safety assessment.

Whereas the self-affirmation process was established in the 1970s, the FDA notification process was introduced in 1997 as an approach to better assure the safety of food ingredients and to reduce the data review burden of the FDA. The Agency retained the authority to deny a GRAS notification due to insufficient or inconsistent safety data. In August 2016, the final ruling on GRAS was posted in the Federal Register replacing the voluntary GRAS affirmation with a voluntary notification procedure (Federal Register, 2016).

Substances which the FDA considered GRAS prior to the 1958 Food Additives Amendment are now listed under 21 CFR § 182. This section includes eight subsections based on the functional characteristic(s) of the substance. Those subsections include: GRAS substances with multiple functions, anticaking agents, chemical preservatives, emulsifying agents, dietary supplements, sequestrates, stabilizers, and nutrients.

Sections 184 and 186 of Title 21 list FDA-affirmed GRAS substances that are appropriate for use in foods without being formally approved as food additives. Section 184 includes GRAS substances that are directly added to foods, such as acetic acid (vinegar) that naturally occurs in plants and produced through fermentation technology, beta-carotene, a nutrient derived from saponification of vitamin A acetate. On the other hand, section 186 lists GRAS substances that are indirectly added to foods, such as dextrans typically produced through bacterial fermentation, and ferric oxide that occurs naturally or may be prepared via specific heating conditions. All of these GRAS substances must be used under GMPs and be used as intended and incorporated under the limits prescribed by the regulation.

7.14 NEW DIETARY INGREDIENT

A new dietary ingredient (NDI) is defined as "a dietary ingredient that was not marketed in the United States before October 15, 1994 (FDCA, 21 U.S.C. 350b(d)). In 2016, the FDA published a draft guidance for the industry which replaced the original guidance issued in July 2011 (Food and Drug, 2018b). According to this document, dietary supplement ingredients in the food supply prior to 1994 are exempt from any requirement for FDA review or approval. The current regulation stipulates the manufacturers or distributors of an NDI must submit a premarket safety notification to the agency at least 75 days prior to the product entering interstate commerce. It is important to note that this requirement is not an FDA approval or clearance. The notification is an opportunity for the Agency to respond should there be any uncertainty as to the safety of the NDI prior to entering the market. Specific requirements for NDI notification are shown in Fig. 7.4.

7.15 INVESTIGATIONAL NEW DRUG

Typically, an investigational new drug (IND) application is submitted to the FDA when a sponsor begins the process of drug development and prepares to initiate studies to examine the safety and therapeutic potential of the isolated and characterized molecule (see Chapter 3) (Food and Drug Administration, 2018b). This regulatory process was extended to clinical studies for conventional foods and dietary supplements to support a new or expanded health claim. However, the IND requirement was stayed on October 30, 2015 (Federal Register, 2015).

7.16 CLEAN LABELS (USA)

According to Innova Market Insights, consumer attitudes have driven the clean label movement. This movement has its foundation in a number of factors: The demand by the consumers for natural foods and food ingredients; mandate for non-GMO foods; distrust in the food industry; and industry's use of food ingredients that consumers cannot pronounce (Innova Market, 2016). In addition, consumers expect clean label products to be more healthful than those foods with artificial or synthetic ingredients. From a regulatory perspective, clean label is not defined. As the food industry responds to the clean label demands, the safety criteria for any and all food additives, color additives, and flavors remain a priority for the FDA and the food industry.

Many within the food industry advanced the clean label movement by publishing lists of unacceptable ingredients. For example, Whole Foods (Whole Foods Clean, 2018) and Panera (Panera's, 2018) posted such lists on their respective websites. These efforts are consistent with consumer demands that include "free-from" labels and "clean" labels. Mintel, a marketing intelligence agency, noted several years ago that 84% of consumers sought more natural, less processed foods, while 43% of consumers thought such products were more healthful. In addition, 59% of consumers called for fewer ingredients on the ingredient declaration, yet only about one-third (37%) of these consumers were willing to pay more for such product changes (Mintel, 2018).

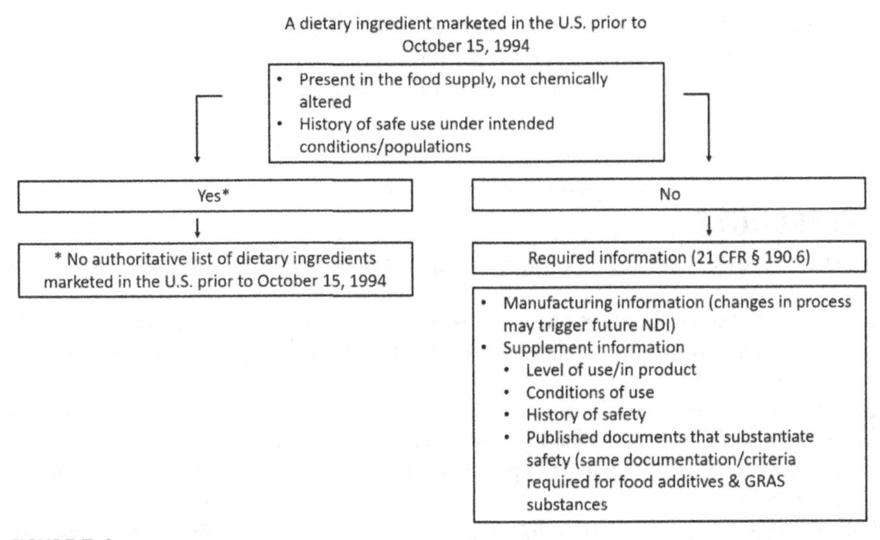

FIGURE 7.4

Requirements for NDI Notification (Based on Dietary Supplements: New DietaryIngredient Notifications and Related Issues: Guidance for Industry, August 2016).

7.17 WHAT IS FREE?

The clean label movement includes consumers seeking a "free-from" content statement in food products, such as gluten-free (Verrill, Zhang, & Kane, 2013). However, under US regulations, a "free" statement does not translate to "zero. " For example, gluten-free means a product contains less than 20 ppm of gluten in a serving (21 CFR § 101.91). Similarly, fat-free milk contains less than 0.5% fat (21 CFR § 131.125), trans-fat free means less than 0.5 mg per serving (21 CFR § 101.62), saturated fat free means less than 0.5 g per labeled serving (21 CFR § 101.62), sugar-free is less than 0.5 g per serving or RACC (21 CFR § 101.60), sodium-free is less than 5 mg per serving or RACC (21 CFR § 101.61), and calorie-free is less than 5 calories (kcal) per serving (21 CFR § 101.60).

7.18 CONCLUSION

The spectrum of food regulations is intended to assure the safety of the food supply chain and to promote the ongoing availability of safe food to the entire population. The establishment of the FDA and the ensuing regulatory agencies with food safety responsibilities has enabled the United States to provide one of the safest food supplies in the global market. A fundamental understanding of food regulations will enable agricultural processes, food producers, and consumers to make better choices for their respective lifestyles. Fundamentally, the food regulations described in this synopsis outline ventures to improve the food supply such that it is safe, nutritious, accessible, and affordable to the entire US population and those countries to which foods are exported.

REFERENCES

AAFCO. (2018). Natural definition. http://talkspetfood.aafco.org/natural. Accessed 01.10.2018.

Arnold, L. E., Lofthouse, N., & Hurt, E. (2012). Artificial food colors and attention-deficit/hyperactivity symptoms: Conclusions to dye for. *Neurotherapeutics*, *9*, 599–609 doi: 10.1007/s13311-012-0133-x.

Bateman, B., Warner, J. O., Hutchinson, E., et al. (2004). The effects of a double blind, placebo controlled, artificial food colourings and benzoate preservative challenge on hyperactivity in a general population sample of preschool children. *Archives of Disease in Childhood*, *89*, 506–511.

California. (2018). Proposition 65. https://oehha.ca.gov/proposition-65. Accessed 01.10.2018.

Centers for Disease Control. (2018). ADHD data and statistics. https://www.cdc.gov/ncbddd/adhd/data.html. Accessed 01.10.2018.

Clydesdale, F., et al. (2005). Functional foods: Opportunities and challenges. Institute of Food Technologists, Expert Report.

Cottage Food Operations. (2016). California. http://www.cdph.ca.gov/programs/Pages/fdbCottageFood.aspx. Accessed 12.12.2016.

Dietary Supplements. (2018). Health and Education Act 1990, Public Law 103-417. https://ods.od.nih.gov/About/DSHEA_Wording.aspx. Accessed 01.10.2018.

Environmental Protection Agency. (2018). https://www.epa.gov/minimum-risk-pesticides/minimum-risk-pesticides-inert-ingredient-and-active-ingredient-eligibility. Accessed 01.10.2018.

FDA. (2016). http://www.fda.gov/Food/GuidanceRegulation/FSMA/default.htm. Accessed 12.12.2016.

FDA Nutrition Facts Panel Changes. (2018). https://www.fda.gov/Food/GuidanceRegulation/GuidanceDocumentsRegulatoryInformation/LabelingNutrition/ucm385663.htm. Accessed 01.10.2018.

Federal Register. (2015). 80, 66907.

Federal Register. (2016). 81(159): 54960–55055. Substances generally recognized as safe. https://www.federalregister.gov/documents/2016/08/17/2016-19164/substances-generally-recognized-as-safe. Accessed 08.17.2016.

Federal Register. (2018) 83(87): 19860-19889. National bioengineered food disclosure standard. https://www.federalregister.gov/documents/2018/05/04/2018-09389/national-bioengineered-food-disclosure-standard

FDA Safety. (2018). Assessment of bioengineered foods, submissions on bioengineered new plant varieties. https://www.fda.gov/Food/IngredientsPackagingLabeling/GEPlants/Submissions/default.htm.

Feingold, B. F. (1975). Hyperkinesis and learning disabilities linked to artificial food flavors and colors. *American Journal of Nursing*, *75*, 797–803.

Food and Drug Administration. (2018a). Redefine "healthy" claim for food labeling. https://www.fda.gov/Food/NewsEvents/ConstituentUpdates/ucm520703.htm. Accessed 01.10.2018.

Food and Drug Administration. (2018b). Guidance for industry: use of the term "healthy" in the labeling of human food products. https://www.fda.gov/Food/GuidanceRegulation/GuidanceDocumentsRegulatoryInformation/ucm521690.htm. Accessed 01.10.2018.

Food and Drug Administration. (2018c). Draft guidance for industry—Dietary supplements: New dietary ingredient notifications and related issues.

Food and Drug Administration. (2018d). Investigational new drug (IND) application. https://www.fda.gov/drugs/developmentapprovalprocess/howdrugsaredevelopedandapproved/approvalapplications/investigationalnewdrugindapplication/default.htm. Accessed 01.10.2018.

Food, Drug and Cosmetic Act. (2018). Section 402 [342], adulterated food. https://www.gpo.gov/fdsys/pkg/USCODE-2010-title21/html/USCODE-2010-title21-chap9-subchapIV-sec342.htm. Accessed 01.10.2018.

Food Labeling Revision. (2016a). Food labeling: Revision of the nutrition and supplement facts labels. Federal Register 81 (103), 33742.

Food Labeling Revision. (2016b). Food labeling: Serving sizes of foods. Federal Register 81 (103), 34000.

Innova Market. (2016). Insights.

Lupton, J. R., Atkinson, S. A., Chang, N., et al. (2014). Exploring the benefits and challenges of establishing a DRI-like process for bioactives. *European Journal of Nutrition*doi: 10.1007/s00394-014-0666-3.

Martirosyan, D. M., & Singh, J. (2015). A new definition of functional food by FFC: What makes a new definition unique? *Functional Foods in Health and Disease*, *5*, 209–223.

Mintel. (2018). http://store.mintel.com/free-from-food-trends-us-may-2015. Accessed 01.10.2018.

McCann, D., Barrett, A., Cooper, A., et al. (2007). Food additives and hyperactive behavior in 3-year-old and 8/9-year-old children in the community: A randomised, double-blinded, placebo-controlled trial. *Lancet*, *370*, 1560–1567.

National Bioengineered. (2016). Food Disclosure Law of 2016, S 764.

National Organic Program. (2018). https://www.ams.usda.gov/about-ams/programs-offices/national-organic-program. Accessed January 01.10.2018.

Panera's. (2018). Panera's no-no list. https://www.panerabread.com/en-us/company/food-policy-no-no-list.html. Accessed 01.10.2018.

Ranard, K. M., Jeon, S., Mohn, E. S., et al. (2017). Dietary guidance for lutein: Consideration for intake recommendations is scientifically supported. *European Journal of Nursing*doi: 10.1007/s00394-017-1580-2.

Roberfroid, M. B. (2000). Defining functional foods. *Functional Foods*, 9–27. doi: 10.1533/9781855736436.1.9.

Rodgers, S. (2016). Minimally processed functional foods: technological and operational pathways. *Journal of Food Science*doi: 10.1111/1750-3841.13422.

Safe and Accurate. (2015). Safe and accurate food labeling act of 2015; HR 1599.

Sato, Y., Itagaki, S., Kurokawa, T., et al. (2011). In vitro and in vivo antioxidant properties of chlorogenic acid and caffeic acid. *International Journal of Pharmaceuticals*, *403*, 136–138 doi: 10 1016/j.ijpharm.2010.09.03.

Sinclair, Upton (1906). *The jungle*. New York, NY: Doubleday, Jabber, and Co.

Statement of Policy. (1992). Foods derived from new plant varieties. Federal Register May 29, 1992. 57 FR 22984.

US. (1999). United States Court of Appeals, District of Columbia Circuit, No. 98-5043, 98-5084, Decided: January 15, 1999.

Verrill, L., Zhang, Y., & Kane, R. (2013). Food label usage and reported difficulty with following a gluten-free diet among individuals in the USA with coeliac disease and those with noncoeliac gluten sensitivity. *Journal of Human Nutrition and Dietetics*, *26*, 479–487. doi: 10.1111/jhn.12032.

Voluntary Labeling. (2015). Indicating whether foods have or have not been derived from genetically engineered Plants. Federal Register, 80 FR 73194, November 24, 2015.

Whole Foods Clean. (2018). https://gocleanlabel.com/whole-foods-clean-label-list/. Accessed 01.10.2018.

CHAPTER

Veterinary products

David A. Dzanis
Regulatory Discretion, Inc., Santa Clarita, CA, United States

In learning to utilize antibiotics for the control of human and animal diseases, the medical and veterinary professions have acquired powerful tools for combating infections and epidemics.

– Selman Waksman

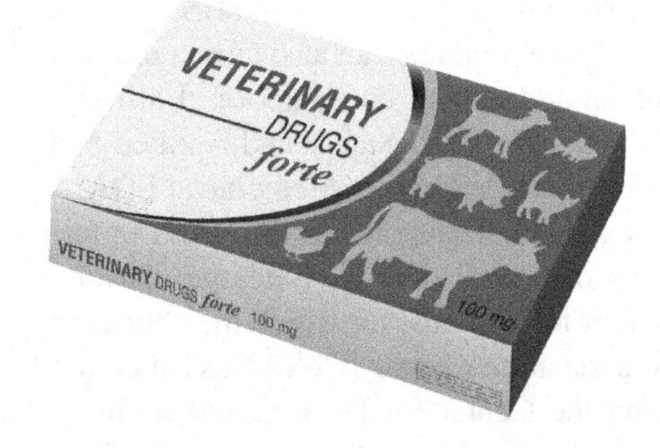

An Overview of FDA Regulated Products. http://dx.doi.org/10.1016/B978-0-12-811155-0.00008-9

CHAPTER OBJECTIVES

After reading this chapter, the reader will be able to:

- describe the differences in oversight of FDA-regulated products intended for animals as compared to the equivalent products for humans,
- identify the unique aspects of the animal drug approval process, particularly for those intended for use in food-producing animals,
- in addition to regulation by FDA, describe the role of state governments in the oversight of animal feeds and pet foods,
- compare regulatory requirements for veterinary devices with those intended for use on humans., and
- identify whether and how FDA regulates animal dietary supplements, veterinary biologics, pesticides for use on animals, and animal grooming aids (akin to "cosmetics" for human use).

8.1 HOW ARE VETERINARY PRODUCTS REGULATED?

Products subject to the Federal Food, Drug, and Cosmetic Act (FFDCA) that are intended for use in animals other than humans are predominantly regulated by the Center for Veterinary Medicine (CVM) within the US Food and Drug Administration (FDA). Use of many of these products can also have a dramatic impact on the safety of foods for both human and animal consumption. Fittingly, CVM works in close association with the Center for Food Safety and Applied Nutrition (CFSAN), both within the Office of Foods and Veterinary Medicine at FDA.

Although probably the smallest center within FDA in terms of staff, CVM performs many of the functions handled by FDA at large. This includes the regulation of foods, drugs, and devices for animals, for which the equivalent products for humans are each handled by a separate Center [CFSAN, Center for Drug Evaluation and Research (CDER), and the Center for Devices and Radiological Health (CDRH), respectively]. However, the functions of the Center for Tobacco Products and the Center for Biologics Evaluation and Research are two areas where CVM does not regulate the animal equivalent.

Within CVM, there are five Offices under the Director (see Fig. 8.1). The primary functions of most of them are explained further below and/or are self-evident. The mission of the Office of Minor Use and Minor Species is similar to that of the Office of Orphan Products for human drugs and devices. Mandated under the Minor Use

FIGURE 8.1 Organization of the center for veterinary medicine.

and Minor Species Animal Health Act of 2004, it identifies and grants special considerations for qualified new animal drugs for either use in minor species (all species other than horses, dogs, cats, cattle, pigs, turkeys, and chickens) or minor uses (for diseases that occur infrequently or in limited geographical areas affecting only a small number of animals annually).

8.2 ANIMAL DRUGS

A drug intended for use in animals must be approved for its intended use in a manner similar to that required for a human drug, that is, via submission of a New Animal Drug Application (NADA) with sufficient data to demonstrate safety and efficacy for its intended use. These are reviewed by the Office of New Animal Drug Evaluation within CVM. Unlike drugs for human use, though, review may include aspects unique to animal drugs, such as an evaluation of residues in the meat, milk, or eggs of animals administered the drug and the subsequent impact these residues may have on food and feed safety. Other functions of CVM not performed by CDER include evaluation of the safety and efficacy of drugs intended purely for economic rather than therapeutic purposes (i.e. "production drugs") and evaluation of drugs intended to be administered via inclusion in the feed of the animal.

Compared to drug testing for human use, the one advantage of conducting drug trials for use in animals is that "preclinical" trials may be conducted on the same species for which the drug is intended. Regardless, animal drugs are also subject to clinical trials (aka "field trials"), same as for human drugs, prior to approval. In other words, the drug must be shown to perform as expected under real-world conditions, not just on animals in the laboratory. Often there needs to be multiple field trials in different geographical locations at different times of the year in order to demonstrate safety and effectiveness under various environmental conditions.

8.2.1 THERAPEUTIC DRUGS FOR NONPRODUCTION ANIMALS

Therapeutic drugs for nonproduction animals include those intended for companion animals, such as dogs, cats, and horses (particular to equine drugs, products must be labeled "Do not use in horses intended for human consumption"). Figs. 8.2 and 8.3 show examples of labeling for therapeutic drugs approved for use in nonproduction animals by FDA. Interestingly, while the FFDCA specifically prohibits any labeling or advertising representation that a new human drug has been approved by FDA, there is no similar prohibition pertaining to labeling or advertising of new animal drugs. Consequently, CVM does allow a statement of fact ("NADA #_____, Approved by FDA.") to be displayed on the labeling or advertising of animal drugs but without further elaboration or undue emphasis.

Sufficient data must be submitted in the NADA to allow CVM to conclude the drug is safe and effective for its intended use. Considerations include life stage of the animal (i.e. use in nonreproducing adults vs. growing animals or those in gestation or

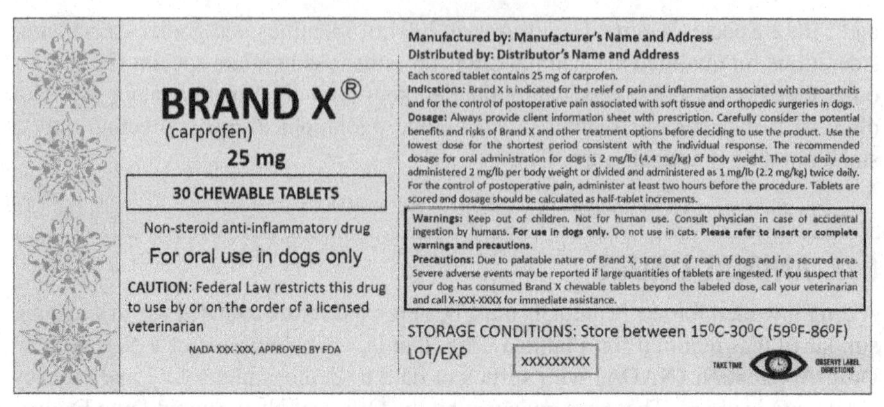

FIGURE 8.2 Example of labeling for a drug approved for use in dogs.

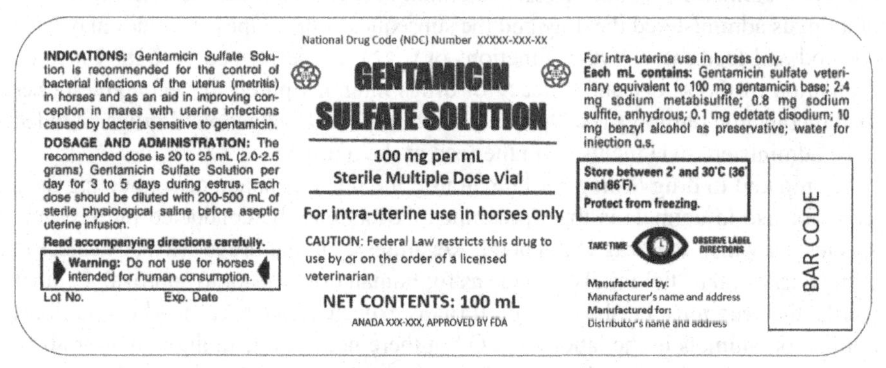

FIGURE 8.3 Example of labeling for a drug approved for use in horses.

lactation), the intended length of treatment (e.g. short-term vs. lifetime), and whether it requires a prescription or can be safely used by the animal owner without any directions from a veterinarian. The drug approval may be restricted to certain life stages, if data are insufficient to demonstrate safety during gestation and lactation, for example. The minimum length of testing for those drugs intended for extended periods if not the lifetime of the animal is 6 months.

8.2.2 THERAPEUTIC DRUGS FOR PRODUCTION ANIMALS

Therapeutic drugs for production animals include those intended for use in cattle, sheep, swine, and poultry. Like for those intended for companion animals, the sponsor must prove animal safety and efficacy for its intended use. Unlike companion animals, the expected life span of the animal may be much shorter (e.g. broiler chickens are typically slaughtered at 5–6 weeks of age). As a result, the length of trials needed to demonstrate safety in the animal may be much shorter. On the other hand, a very significant factor in evaluation not considered for human drugs is the safety of potential residues in the edible tissues of those animals receiving the drug. Extensive studies must be submitted in order for CVM to evaluate the metabolism, deposition, degradation, and excretion of the drug, so as to be able to assess the potential for residues and potential public and animal health concerns. A lot of the work in the Office of Research in CVM is dedicated to the understanding of metabolism of drugs and development of laboratory methods to detect residues. If necessary, additional restrictions, such as an established withdrawal period prior to slaughter, may be imposed on the drug in order to ensure that the edible tissues derived from the treated animals are safe for consumption. An example of labeling for a drug approved for therapeutic use in production animals is shown in Fig. 8.4.

Of particular concern to CVM is the safety of antibiotics and the potential for development of antimicrobial resistance. While the development of resistance may be due to a number of factors, CVM is engaged in multiple activities to reduce the risk stemming from use of animal antibiotics, particularly with respect to use in food-producing animals. New antibiotics are subject to rigorous review with regard to demonstration of human food safety, while previous approvals for some antibiotics have been revoked. NADAs for novel classes of antibiotics may be simply rejected from further consideration for use in animals by CVM, in order to help prevent the future development of antimicrobial resistance to medically important human drugs.

8.2.3 PRODUCTION DRUGS

A "production drug," unlike most drugs for human use, is not intended to cure, mitigate, treat, prevent, or otherwise affect a disease or condition. Rather, it is expressly intended to affect the animal's production, such as to provide an economic advantage in terms of meat, milk, or egg production. Common production drugs include anabolic steroids, which are often administered to beef cattle to improve feed efficiency (weight gain per pound of feed consumed). An example of a label for an approved anabolic steroid combination is shown in Fig. 8.5. Typically, they are in the form

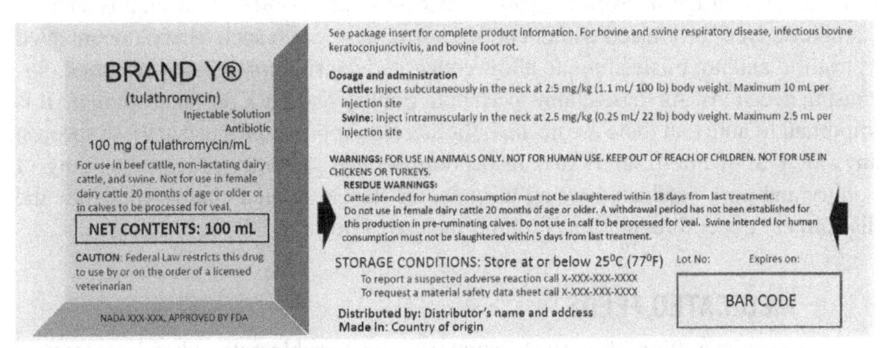

FIGURE 8.4 Example of labeling for a therapeutic drug approved for use in cattle and swine.

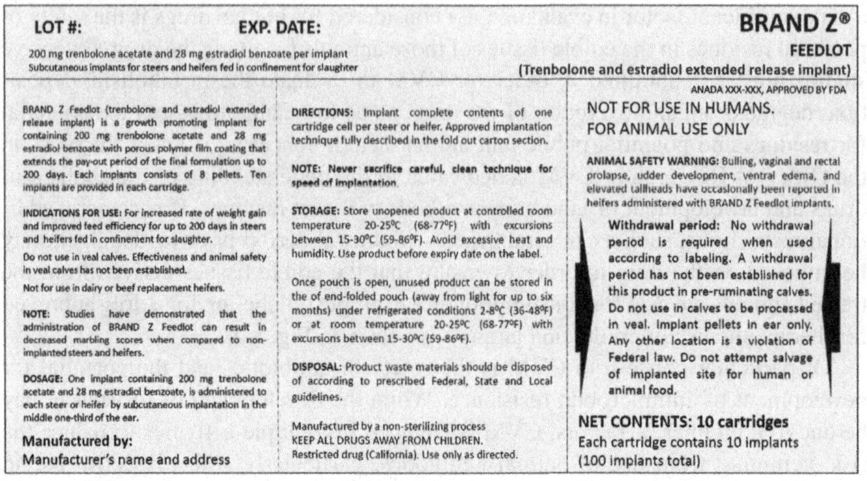

FIGURE 8.5 Example of labeling for an approved production drug in beef cattle.

of impregnated plastic pellets injected subcutaneously. The pellets slowly release the steroids during the production period of the animal, improving weight gain (preferentially, muscle weight) while eating the same amount of food as a nontreated animal. In this way, even when the costs of the drug are considered, the cattle owner can produce the same amount of meat for less cost compared to the one who did not use the drug. Other examples include the injection of bovine growth hormone (BGH, aka bovine somatotropin, BST) to increase milk production in lactating dairy cattle. Also, in addition to effects on disease, many antibiotics have been shown to affect the feed efficiency of the animal when fed at subtherapeutic levels. Historically, a number of antibiotics were approved for these purposes, but under CVM's efforts to curtail microbial resistance these uses are being phased out.

Food labels are not required to declare the possible presence of residues from use of production drugs. However, many human food labels often bear claims such as "antibiotic-free" or "raised without added hormones" when such can be documented as truthful and not misleading, to allow concerned consumers to make informed purchasing decisions. As these claims pertain to poultry and pork products, though, it is important to note that there are no anabolic steroids approved for use in these species, anyway. USDA Food Safety and Inspection Service guidance requires labeling of poultry and pork products bearing hormone or steroid-related claims to include the disclaimer "Federal regulations prohibit the use of hormones."

8.2.4 MEDICATED FEEDS

Efficient administration of a drug to animals can be problematic when large groups of animals must be treated at the same time, such as for an entire herd of cattle or flock of chickens. Thus, a common means of administration is by inclusion of the drug in the animal's feed. "Medicated feeds" are those feeds containing one or more approved drugs. These may be intended for either therapeutic or production purposes, though as mentioned above, the use of antibiotics for production purposes is being phased out in the United States. An example of a label for a medicated feed containing a therapeutic drug is shown in Fig. 8.6.

The approval of the drug to be used in a medicated feed is contingent on proof that the drug is readily available in a feed matrix and effective when delivered by this means. Feed mills incorporating drugs into feed must be licensed and follow strict good manufacturing practices to ensure the drug is in the feed in proper amounts and the feed is sufficiently mixed to achieve uniform distribution. Many medicated feeds, particularly those containing antibiotics, can only be sold to those in possession of a Veterinary Feed Directive. In other words, akin to a prescription, a licensed veterinarian must sign off on the need for use of the medicated feed to address a specific medical condition in a particular group of animals before the feed is dispensed to the animal owner.

In addition to CVM oversight of animal drugs in interstate commerce, some states have also implemented their own "animal remedy" laws. These generally require manufacturers of animal drugs, as well as veterinary biologics, to be licensed in the state and/or register products prior to distribution.

8.3 ANIMAL FEEDS

All animal feeds in interstate commerce are subject to oversight by the Division of Animal Feeds within the Office of Surveillance and Compliance in CVM. Products covered include both livestock feeds and pet foods, which further includes complete and balanced foods as well as treats, nutritional supplements and edible chews (e.g. rawhides), as well as the ingredients intended to be used in these feeds. Although meat and poultry products intended for human consumption are regulated by the Food Safety and Inspection Service in the US Department of Agriculture, once designated for animal feed use these items are solely under CVM control.

8.3.1 FEDERAL VERSUS STATE REGULATORY OVERSIGHT

Although animal feed facilities must be registered with FDA under the Bioterrorism Act, there is no premarket review or registration of animal feed products by CVM. Regardless, the majority of states in the United States administer and enforce their own feed regulations, many of which require company licensure, product registration, or both as a condition of distribution of the animal feed product within the jurisdiction of that state. This generally necessitates submission of labeling to the state feed control official (typically, but not always in that state's department of agriculture) for review prior to approval. While one state's denial of a product for distribution does not affect acceptance in other states, it is typically infeasible for a manufacturer to print different labels to abide by each state's requirements, so a company intending to distribute

BLUE BIRD LABELING
bacitracin methylenedisalicylate
Type B medicated feed for beef steers & heifers fed in confinement for slaughter – Liver Abscess 70
(bacitracin Type B medicated feed)

Do not Feed Undiluted

For reduction in the number of liver condemnations due to abscesses in beef steers and heifers fed in confinement for slaughter.

Active Drug Ingredient
Bacitracin (as feed grade bacitracin methylenedisalicylate)* .... 51 – 50,000 g/ton(.025 – 25 g/lb)**

Guaranteed Analysis
Crude Protein (Min)..______%
NPN[a] (Max)..______%
Crude Fat (Min)..______%
Crude Fiber (Max)..______%
Calcium[a] (Min)..______%
Calcium[a] (Max)..______%
Phosphorus[a] (Min)..______%
Salt[a] (Min)..______%
Salt[a] (Max)..______%
Sodium[b] (Min)..______%
Sodium[b] (Max)..______%
Potassium[a] (Min)..______%
Vitamin A[a] ....______IU/lb

[a] Guarantee required only when nutrient added except when the feed is intended, represented or serves as a principal source of the nutrient.
[b] Sodium guarantee required only when total sodium exceeds that furnished by the maximum salt guarantee.

Ingredients
Feed ingredients, as defined by AAFCO, must be listed on the final printed Type C medicated feed label by their common or usual names in descending order or predominance by weight.

Mixing Directions
Mix this Type B medicated feed with non-medicated feed ingredients to manufacture one ton of cattle feed.
The following table provides examples of mixing rates:

Type B bacitracin concentration (g/ton)	Type B per ton of Type C	Non-medicated feed per ton of Type C (lb)	Type C bacitracin concentration (g/ton)
100 (0.05 g/lb)	100	1900	5
1000 (0.5 g/lb)	10	1990	5
100 (0.05 g/lb)	200	1800	10
1000 (0.5 g/lb)	20	1980	10

The resulting Type C medicated feed should be administered continuously throughout the feeding period to provide 70 mg bacitracin per head per day.

Manufactured By:
Blue Bird Feed Mill
City, State, Zip

NET WEIGHT ON BAG OR BULK

Lot # ______________

*Pennitracin MD is the proprietary name of bacitracin methylenedisalicylate formulation (Type A Medicated Article).
**The final printed feed label should list only a single drug concentration.

BLUE BIRD LABELING

Version 9/10/15

FIGURE 8.6 Example of labeling for a medicated feed containing an approved drug for cattle.

in multiple states will generally construct its labels to meet all of those applicable. Thus, unlike regulation of most foods for human consumption under CFSAN, there effectively is a nation-wide premarket review of labeling for animal feeds.

Clearly, attempting to construct a label to meet up to 50 different sets of regulations can be daunting. Fortunately, the Association of American Feed Control Officials (AAFCO) was formed in 1909 with the mission of encouraging uniform interpretation and enforcement of animal feed laws between the states. AAFCO is a nongovernmental body, but whose members must be representatives of state and territorial agencies charged with the regulation of animal feeds. Federal agencies, such as CVM, as well as foreign governments (Canada, Costa Rica) are also represented within AAFCO.

Federals regulations governing the labeling of animal feeds are few (21 CFR 501). They cover the basic requirements for a statement of identity, a net content declaration, an ingredient declaration, and the manufacturer's or distributor's name and address. They do *not* address many of the other aspects of labeling as required by CFSAN for human foods under 21 CFR 101, though, most notably, CVM regulations do not require the equivalent to a Nutrition Facts Box, nor any other declarations in regard to nutrient content on animal feed labels at this time. However, the AAFCO Model Bill and Regulations serve to fill this gap.

The AAFCO Model Bill and Regulations constitute a consensus among its members as to the appropriate regulation of feed. These documents are continuously debated and revised as needed to address new or evolving issues. Importantly, AAFCO has no authority to enforce these models itself. However, states are free to adopt these models for incorporation into their own statutes, where they do have the power to enforce the rules within their respective jurisdictions. A majority of states have adopted the AAFCO models in some form, and while each state can only take action on violative products within its borders, by acting in concert with other states effective oversight is achieved nationwide. States and CVM interact frequently on many enforcement matters, so feed control officials become aware of the discovery of a noncompliant product in one state and will seek action against it within their jurisdictions as well.

Another important aspect of AAFCO is the establishment of standardized definitions and terms for feed ingredients. FDA (both CFSAN and CVM) regulations only stipulate use of the "common or usual" name of the ingredient in the ingredient declaration. While that may apply to many animal feed (particularly pet food) ingredients, many other ingredients are unique to animal feed. To avoid confusion between jurisdictions, then, ingredients must conform to the AAFCO Official Common and Usual Names and Definitions of Feed Ingredients and be declared on the label by that name. The exception is to allow use of "collective terms" for livestock and poultry feeds (but not pet foods). For example, a feed mill may replace wheat for corn in a sheep ration, depending on availability and price of the commodity. In either case, though, it appears on the label as "grain products."

Similar to CFSAN's function with regard to human food ingredients, CVM reviews food additive petitions (FAPs) and GRAS (generally recognized as safe) notifications for ingredients specifically intended to be incorporated in animal feed (21 CFR 570–584). However, many ingredients are also deemed acceptable for use in animal feeds via the AAFCO feed ingredient definition petition process. Arguably a less burdensome process than either an FAP or a GRAS notification, the quantity and quality of data required to show safety and utility of a new animal feed ingredient through AAFCO can be extensive. Although AAFCO facilitates the process, CVM is responsible for the review of all safety and utility data associated with that new ingredient. Only after CVM signs off on the acceptability of the ingredient can AAFCO proceed with publication of the new or amended definition.

8.3.2 LABELING REQUIREMENTS FOR ANIMAL FEEDS

The labels of feeds intended for livestock and poultry are often bare-boned, containing only the information as mandated by either CVM or AAFCO regulations. These are often printed on a tag that is sewn to one end of the feed bag. Labeling requirements include:

- product name (and brand name, if any),
- purpose statement (intended species and class),
- guaranteed analysis,
- ingredient declaration,
- directions for use and precautionary statements, if any,
- name and address of the manufacturer or the distributor, and
- quantity statement.

An example of an acceptable feed label can be seen in Fig. 8.7. The "guaranteed analysis" is a declaration of nutrient content, similar in intent to the Nutrition Facts Box. While the formats are quite different, both are designed to convey important nutritional information to the purchaser.

8.3.3 LABELING REQUIREMENTS FOR PET FOODS

Pets include dog and cats, as well as "specialty pets" (e.g. reptiles and amphibians, rodents and other small mammals, pet birds, and ornamental fish). Compared to livestock feed labels, labels of products for these species are generally much more extensive. They need to address additional aspects of nutrition not covered in the general regulations. Also, pet foods are often given fanciful names, elaborate on inclusion of specific ingredients, and can bear considerable amounts of "romance copy" in addition to the mandatory verbiage. Thus, AAFCO has also promulgated the Model Regulations for Pet Food and Specialty Pet Food to specifically address these unique aspects of the labeling.

Additional mandatory information on dog and cat food versus livestock feed labels include a calorie content statement and a nutritional adequacy statement (see Fig. 8.8). Calories must be expressed both in terms of kilocalories by weight (kilograms) and per familiar household measure (such as cups, cans, or pieces). Calories

BLUE BIRD LAYER FEED

Complete Feed for Laying Chickens

Guaranteed Analysis

Crude Protein (Min)	15.0%
Lysine (Min)	**0.65%**
Methionine (Min)	**0.35%**
Crude Fat (Min)	3.0%
Crude Fiber (Max)	3.5%
Calcium (Min)	**3.4%**
Calcium (Max)	**4.4%**
Phosphorus (Min)	**0.6%**
Salt (Min)	**0.35%**
Salt (Max)	**0.50%**

Ingredient Statement

Grain Products, Plant Protein Products, Processed Grain By-Products, Animal Protein Products, Vitamin A Supplement, Vitamin D3 Supplement, Vitamin E Supplement, Riboflavin Supplement, Niacin Supplement, Calcium Pantothenate, Choline Chloride, Folic Acid, Menadione Sodium Bisulfite Complex (Source of Vitamin K Activity), Methionine Supplement, Potassium Sulfate, Calcium Carbonate, Salt, Manganous Oxide, Ferrous Sulfate, Copper Sulfate, Zinc Oxide, Ethylenediamine Dihydriodide, Sodium Selenite.

Feeding Directions:

This is a complete and balanced ration. Feed Blue Bird Layer Feed from time first egg is laid throughout the time of egg production. Always provide plenty of fresh water.

Manufactured By:
Blue Bird Feed Mill
City, State Zip

NET WT 50 LB (22.67 kg)

FIGURE 8.7 Example of an acceptable poultry feed label.

(Courtesy of AAFCO).

are most often determined by a calculation method using the "modified Atwater" values. These values approximate the average calorific content of the protein, fat, and carbohydrate fractions of the food, but are different from the Atwater values used for the same determinations for human food. Alternatively, calories may be determined through conduct of a digestibility trial on dogs or cats, as appropriate, to estimate metabolizable energy content of the food.

Because many dog and cat foods are intended to be "complete and balanced," rigorous standards are in place to help ensure the product could in fact serve as the sole source of nutrition (except water). Means of substantiation of nutritional adequacy include formulation to meet the established minimum and maximum concentrations of all recognized essential nutrients as stipulated in the AAFCO Dog and Cat Food Nutrient Profiles. Different profiles exist for each species and for different life stages (adult maintenance vs. growth and reproduction). Those meeting the requirements for both may be claimed for "all life stages."

Alternatively, nutritional adequacy can be substantiated via successful passage in a feeding trial following AAFCO protocols. A specified number of animals must be fed the test diet as the sole source of nutrition for a specified period of time. Animals are subject to evaluation by physical examination, measurement of blood parameters and serum chemistries, and depending on intended life stage, other measurements (e.g. growth of puppies or kittens, litter size and survivability in gestating/lactating females). These parameters are judged against those from animals in a concurrent or historical control group that were fed a diet that previously passed the feeding trial. To be suitable for "all life stages," puppies or kittens obtained from dams that successfully passed the gestation/lactation trial must be fed the same food throughout an immediately

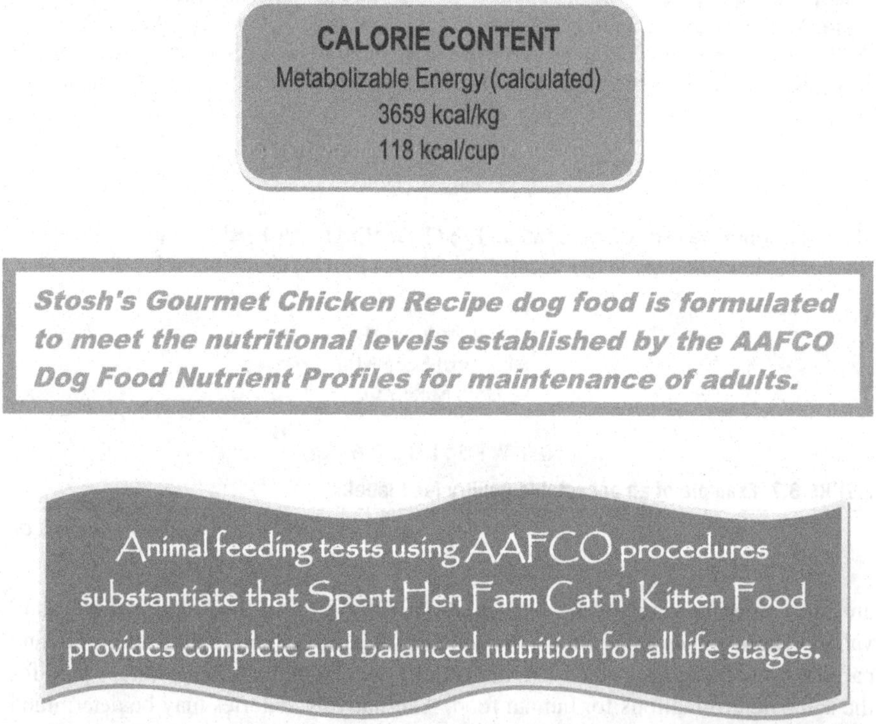

CALORIE CONTENT
Metabolizable Energy (calculated)
3659 kcal/kg
118 kcal/cup

Stosh's Gourmet Chicken Recipe dog food is formulated to meet the nutritional levels established by the AAFCO Dog Food Nutrient Profiles for maintenance of adults.

Animal feeding tests using AAFCO procedures substantiate that Spent Hen Farm Cat n' Kitten Food provides complete and balanced nutrition for all life stages.

FIGURE 8.8 Examples of acceptable calorie content and nutritional adequacy statements on a pet food label.

subsequent growth trial. If those puppies or kittens also pass, the nutritional suitability to meet the less demanding needs of nonreproducing adults is assumed.

In either case, an appropriate statement indicating the method of substantiation and intended life stage must appear on the label (see Fig. 8.8). The only exception is for products not intended to provide complete and balanced nutrition, such as those labeled prominently as "treats," "snacks," or "supplements." Those that are not labeled as such but do not meet either method must state "This product is intended for intermittent or supplemental feeding only." Although far less known and understood by the public, AAFCO also currently provides for a third method of substantiation, where a product can be considered complete and balanced if it is part of a "product family." This requires the product meet certain nutritional criteria relative to a product that did pass a feeding trial, but does not wholly account for other factors such as bioavailability of nutrients from different ingredients. Unfortunately, the label statement for a product family member may appear identical to that for a product that was actually the subject of a feeding trial.

Pet food companies often use names of ingredients as a constituent of the product variety name. So as not to perpetuate a false impression as to the amount of the named ingredient in the product, AAFCO regulations stipulate minimum percentages of ingredients necessary to support a name. For example, "Chicken," "Chicken Recipe," and "With Chicken" requires that the product contain at least 95%, 25%, or 3% chicken, respectively. Those containing less than 3% of the named ingredient can only be designated as a "flavor." Also, except for a "flavor," the name of the ingredient in the product variety name must match that in the ingredient list. It would be false and misleading, for example, for a product to be labeled "Chicken Formula" in accordance with the 25% rule, for example, if in fact the ingredient was not "chicken" (flesh and skin, with or without bone) but rather a form of chicken by-products (e.g. chicken heads, feet, viscera, or a rendered meal from carcasses).

Unlike most livestock feeds, which are typically intended to support body weight if not increase growth of the animal consuming it, the prevalence of overweight or obese dogs and cats in the United States has prompted the formulation of products expressly intended for weight control or reduction. Claims for "low calorie," "lite, "low fat," etc. must meet specific maximum calorie or fat restrictions depending on whether for a dog or cat, and the type of product (dry, semi-moist, or canned). Those not meeting those criteria but still relatively lower than another product may still make a comparative claim (e.g. "X% less calories than Y").

8.4 HOW ARE DIETARY SUPPLEMENTS FOR ANIMALS REGULATED?

CVM and state feed control officials also regulate the distribution of nutritional supplements for animals. The Dietary Supplement Health and Education Act of 1994 (DSHEA) provides for the use of ingredients in products for human consumption meeting the statutory definition of a dietary supplement that are not necessarily

GRAS, approved food additives, or otherwise sanctioned for use in food in conventional form. However, in consideration of the inferred intent of Congress in passing this law, CVM has determined that DSHEA does *not* apply to products for animal use. As a result, animal supplements may contain nutrients such as vitamins, minerals, fatty acids, etc., but not many of the herbs, metabolites, and other compounds frequently found in the dietary supplement aisle at most retail outlets. Also, contrary to labels of human dietary supplements, animal supplement labels cannot bear "structure/function" claims, unless that claim is directly related to its established nutritive value. For example, "with calcium to support bone health" is perfectly acceptable on an animal supplement label, because calcium is understood to be an essential nutrient, that is, required in the diet for normal bone structure and function. However, "with garlic to support a healthy immune system" would not be acceptable, because the recognized function of garlic in animal feed is to provide taste and aroma, not for the purpose of affecting immune health.

Notwithstanding the regulatory status of dietary supplements for animal use, many products containing unapproved feed ingredients do exist in the market. Many, particularly those intended for dogs, cats, or horses, are labeled per the guidance from the National Animal Supplement Council (NASC), a trade organization. Labeling of these products do not follow CVM/AAFCO requirements (e.g. no guaranteed analysis or ingredient list as would be required on an animal feed or pet food label), but rather bear a "Products Facts Box," which is very similar to a Supplement Facts Box as appears on human products labeled as per DSHEA (see Fig. 8.9). They are not represented as foods or as containing ingredients of any nutritive value, but rather described as "dosage-form animal health products." In this way, they escape scrutiny by most state feed control officials because they are not "feeds." CVM and states enforcing animal remedy laws consider them "unapproved drugs of low regulatory priority," but they are generally tolerated under enforcement discretion.

8.5 VETERINARY DEVICES

Because the pertinent sections of FFDCA in relation to devices refer to articles intended for "man or other animals," items such as surgical instruments, needles and syringes, prosthetic devices, certain diagnostic test kits, and X-ray equipment intended for animals meet the statutory definition of "devices" [21 USC 321(h)]. In many cases, the equipment utilized by the veterinarian may be exactly the same as that employed by the human physician, although in quite a few circumstances devices are specially made for the practice of veterinary medicine to accommodate differences in anatomy and intended uses. Regardless, CVM does have regulatory oversight of veterinary devices and can take appropriate enforcement action via its Office of Surveillance and Compliance against products on the market found to be adulterated or misbranded. It is the burden of the manufacturer and/or distributor to ensure that products on the market are safe, effective, and properly labeled. Manufacturers and distributors are urged to seek a review of labeling and promotional materials for veterinary devices to help ensure that products are in compliance with FFDCA.

Melanie's Happy Hoof Co.

HEALTHY JOINT SUPPORT FOR HORSES

Product Facts

ACTIVE INGREDIENTS PER OUNCE:

Glucosamine HCl (shellfish)	1500 mg
Chondroitin Sulfate (poultry)	800 mg
Proprietary Herbal Blend (Ginger Root, Boswellia, Turmeric)	250 mg
Ascorbic Acid (Vitamin C)	100 mg
Manganese (manganese amino acid complex)	15 mg

INACTIVE INGREDIENTS:
Artificial apple flavor, beet molasses, BHA and BHT (preservatives), dehydrated alfalfa meal, oat hulls, propionic acid (preservative) and silicon dioxide

CAUTIONS:
Safe use in pregnant animals or animals intended for breeding has not been proven.
If lameness worsens, discontinue use and contact your veterinarian.
Administer during or after the animal has eaten to reduce incidence of gastrointestinal upset.

FOR USE IN HORSES ONLY.
RECOMMENDED TO SUPPORT HEALTHY JOINT FUNCTION.
DIRECTIONS FOR USE:
Enclosed scoop holds 1 ounce.
Give 2 scoops morning and night for initial loading dose (3-4 weeks).
For maintenance, give 1 scoop daily (based on an 1000 lb adult horse).

WARNINGS:
For animal use only.
Keep out of the reach of children and animals. In case of accidental overdose, contact a health professional immediately.
This product should not be given to animals intended for human consumption.

QUESTIONS?

Distributed by: Melanie's Happy Hoof Co., San Angeles, CA 90298

Call us at **1-800-555-4959** or visit **www.melanieshappyhoof.com**

Lot #: *Best before:*

FIGURE 8.9 Example of a "dosage-form animal health" (aka "supplement") product for horses labeled as per NASC.

On the other hand, other aspects of the law pertaining to devices do not apply to those specifically designed for veterinary use. Very importantly, devices distributed exclusively for use in animals are not subject to any form of premarket approval. Furthermore, manufacturers who distribute devices for veterinary use only are not required to register their establishments, and such devices are exempt from postmarketing reporting as required for devices intended for humans. The exception to this exemption for veterinary products is for firms that manufacture radiation-emitting devices. They do need to register their products under the regulations administered by the Center for Devices and Radiological Health.

Many pet chews and treat labels bear claims relating to dental health and hygiene. Provided that the product achieves this effect by mechanical and not chemical means, they are most often construed as device claims. For example, a hard, edible item that allows for prolonged chewing and contact time with the mouth may bear a claim such as "helps scrape away dental plaque" or "massages gums" without making the product a drug. These types of claims for products on the market cannot be false or misleading, though the products themselves are not subject to premarket review by CVM. However, state feed control officials may still review them as food claims prior to any manufacturer licensure or product registration.

8.6 VETERINARY BIOLOGICS

One category of products overseen by FDA for human use but that CVM does not regulate is biologics (e.g. vaccines, blood products, diagnostic kits). Rather, these items are subject to oversight by the Animal and Plant Health Inspection Service (APHIS) within the US Department of Agriculture. Its jurisdiction over this category of products is authorized by the Virus-Serum-Toxin Act [21 USC 151–159].

Every biologic manufactured in the United States must hold a Veterinary Biological Product License issued by the Center for Veterinary Biologics in APHIS. Successful licensure requires demonstration that the product is pure, safe, potent, and effective. Manufacturers of veterinary biologics in the United States also must have a Veterinary Biologics Establishment License. A manufacturer must hold at least one product license in order to qualify for the establishment license. For those companies wishing to market imported veterinary biologics in the United States, a Veterinary Biological Product Permit (permit for distribution and sale) through APHIS is required.

Biologics such as blood products and colostrum are also acceptable as animal feed ingredients as defined by AAFCO. However, labels for feeds containing these ingredients cannot make any expressed or implied claims regarding immune modulation or other effect characteristic of a biologic. Rather, claims are restricted to those that can be ascribed to food (taste, aroma, or nutritive value).

8.7 PESTICIDES

An item intended to affect external parasites of animals such as fleas and ticks may be subject to regulation as a drug, as a pesticide, or both. While CVM has jurisdiction over animal drugs, pesticides are regulated by the Environmental Protection Agency (EPA) under the authority of the Federal Insecticide, Fungicide, and Rodenticide Act

(FIFRA). The determination of which agency has primary jurisdiction over a given product is laid out in a memorandum of understanding between the agencies.

Generally, EPA has jurisdiction over agents applied topically on the animal for direct effect or to the environment of the animal (e.g. lawns, bedding) and will register them as pesticides. However, although there are exceptions, those parasite-control agents that work systemically, such as those administered orally or parenterally to the animal, are considered drugs and may require CVM approval. This category includes agents that may be applied topically, but that function only after absorption through the skin and dispersal of the agent through the systemic circulation.

Where dual authorities exist, the two agencies will interact. The primary agency may be determined by the types of claims for the product, that is, whether the principal representations relate more to its function as a pesticide or as a drug. A product that is subject to both FIFRA and FFDCA may not be registered by EPA until it has been notified by CVM that it also is in compliance with any requirements set forth as an animal drug under FFDCA. Similarly, CVM may not approve a new animal drug application prior to notification by EPA that the product abides by FIFRA and is also eligible for registration as a pesticide.

Many states also administer their own programs related to the sale, application, transport, and disposal of pesticides. In fact, many of the state feed control officials mentioned above are in charge of these efforts as well (as well as state programs regarding regulation of fertilizers). The Association of American Pesticide Control Officials (AAPCO) was formed in 1947 to facilitate communication and interaction with federal agencies under FIFRA.

8.8 ANIMAL GROOMING AIDS

The term "cosmetic" as defined by FFDCA applies specifically to articles applied to the "human body," hence does not apply to animal products. Of course, there are products on the market intended for cosmetic purposes in animals, that is, for cleansing, beautifying, promoting attractiveness, or altering the appearance of animals rather than human beings. These are classified as "grooming aids" by CVM.

Animal grooming aids are not normally subject to CVM review or enforcement action, provided the labeling for such products do not indicate or imply therapeutic use. For example, a shampoo labeled to promote the luster of the fur of dogs and cats is not subject to regulation, while one containing an "active" ingredient intended for management of seborrhea, or even just a label claim to "help fight itchy, scaly skin" could be subject to regulation as a drug. Similarly, a flea and tick shampoo or rinse would be subject to regulation as a pesticide by EPA.

Absence of regulatory oversight also presumes that the product is intended exclusively for use in animals. Those products intended for both humans and animals would otherwise need to comply with any pertinent requirements for cosmetics under FFDCA. This would include the need for the label to bear appropriate use directions for both humans and animals.

As mentioned above, many pet chew and treat labels may bear claims relative to dental health. When claims are limited to appearance or attractiveness only (e.g. "to help clean teeth" or "freshens breath") by virtue of mechanical action or inclusion of herbs such as mint or parsley to mask odors, these are often construed as cosmetic claims by CVM and are not actionable. Again, though, state feed control officials may require modification of a claim deemed to be false or misleading as a condition of distribution in the state as an animal feed.

FURTHER READINGS

AAFCO. (2017). *Official publication*. Champaign, IL: Association of American Feed Control Officials, Inc.

Animal and Veterinary. (2017). http://www.fda.gov/cvm Accessed 08.04.17.

Business of Pet Food. (2017). http://www.petfood.aafco.org/ Accessed 08.04.17.

GPO. (2017). *Code of Federal Regulations, Title 21, Parts 500–599*. Washington, DC: Government Printing Office.

Pesticide Registration. (2017). https://www.epa.gov/pesticide-registration Accessed 08.04.17.

Veterinary Biologics. (2017). https://www.aphis.usda.gov/aphis/ourfocus/animalhealth/veterinary-biologics Accessed 08.04.17.

CHAPTER

Dietary supplements

Anthony L. Young*, James William Woodlee*, Michael M. McGuffin**

**Kleinfeld, Kaplan and Becker LLP, Washington, DC, United States;*

***American Herbal Products Association, Silver Spring, MD, United States*

"Millions of Americans today are taking dietary supplements, practicing yoga and integrating other natural therapies into their lives. These are all preventive measures that will keep them out of the doctor's office and drive down the costs of treating serious problems like heart disease and diabetes."

– Andrew Weil

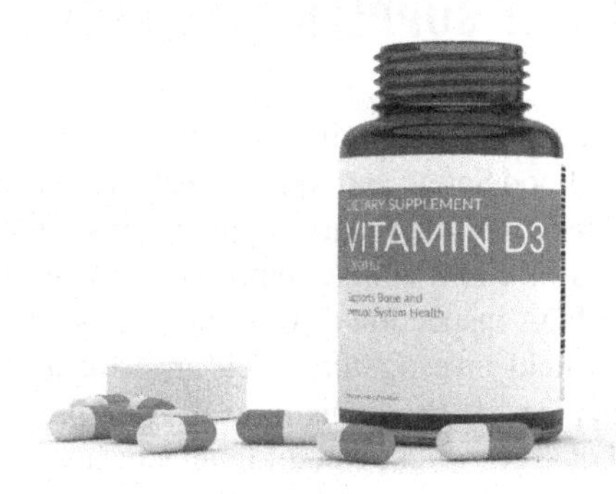

An Overview of FDA Regulated Products. http://dx.doi.org/10.1016/B978-0-12-811155-0.00009-0

CHAPTER OBJECTIVES

After reading this chapter, the reader will be able to:

- define the "dietary supplements" category of products as compared to foods and drugs and understand the limitations on ingredients that are allowed in these products and the claims that may be made for them,
- understand the definition of new dietary ingredients and the available pathways for bringing new dietary ingredients to market,
- understand the requirements for labeling dietary supplements and the claims that can be made for them,
- identify the current good manufacturing requirements that apply to dietary supplements and those provisions of the Bioterrorism Act of 2002 and the Food Safety Modernization Act that apply to this class of foods,
- describe how state and local enforcement authorities and consumer legal actions oversee and enforce against unlawful dietary supplements, and
- understand that while dietary supplements are regulated by the FDA and its regulatory scheme, they are, like foods, and unlike drugs and medical devices generally, a category that does not require FDA premarket approval.

9.1 HOW ARE DIETARY SUPPLEMENTS REGULATED WITHIN FDA?

Dietary supplements are regulated mainly by the Office of Dietary Supplement Programs under the Center for Food Safety and Applied Nutrition located in FDA's Office of Food and Veterinary Medicine. As dietary supplements do not require premarket approval, a manufacturer's first real contact with the FDA most likely takes place when it is subjected to an FDA inspection or web site claims review by the Office of Human and Animal Food field investigators. Field investigators work closely with the Office of Dietary Supplement Programs in evaluating product labels, labeling, and web site claims, and dietary current Good Manufacturing Practices compliance.

9.2 WHAT ARE DIETARY SUPPLEMENTS?

In 1994, the Dietary Supplement Health and Education Act (DSHEA) defined and created the category of food known as dietary supplements (Public Law 108-417, 1994):

According to 21 USC 321(ff), "dietary supplement"—

(1) means a product (other than tobacco) intended to supplement the diet that bears or contains one or more of the following dietary ingredients:(A)a vitamin;(B)a mineral;(C)a herb or other botanical;(D)an amino acid;(E)a dietary substance for use by man to supplement the diet by increasing the

total dietary intake; or(F)a concentrate, metabolite, constituent, extract, or combination of any ingredient described in clause (A), (B), (C), (D), or (E).

(2) *does—(A)include an article that is approved as a new drug under Section 355 of this title or licensed as a biologic under Section 262 of title 42 and was, prior to such approval, certification, or license, marketed as a dietary supplement or as a food unless the Secretary has issued a regulation, after notice and comment, finding that the article, when used as or in a dietary supplement under the conditions of use and dosages set forth in the labeling for such dietary supplement, is unlawful under Section 342(f) of this title; and(B)not include—(i)an article that is approved as a new drug under Section 355 of this title, certified as an antibiotic under Section 357 of this title, or licensed as a biologic under Section 262 of title 42, or(ii)an article authorized for investigation as a new drug, antibiotic, or biological for which substantial clinical investigations have been instituted and for which the existence of such investigations has been made public, which was not before such approval, certification, licensing, or authorization marketed as a dietary supplement or as a food unless the Secretary, in the Secretary's discretion, has issued a regulation, after notice and comment, finding that the article would be lawful under this chapter.*

This list of dietary ingredients goes from a specific nutrient component, for example, a vitamin, to the broad description of "dietary substance for use by man to supplement the diet by increasing the total dietary intake," which could include anything that has been consumed. The FDA has, however, interpreted "dietary substance" as a substance that already is or has been in the diet, and not one that is, for the first time, created to be added to the diet. The law further defines dietary supplement as intended for ingestion and not represented for use as a conventional food or as a sole item of a meal or the diet and to be labeled as a dietary supplement.Early litigation after DSHEA firmly established that dietary supplements are for ingestion through the mouth and not through any other orifice (United States v. Ten Cartons Ener-B Nasal Gel, 888 F. Supp. 381, 393-94 (E.D.N.Y.), aff'd, 72 F.3d 285 (2d Cir. 1995)). Hence, dietary supplements may be in capsule, tablet, liquid, powder, soft gel, gelatin capsule, or gummy form or in a conventional food form (such as beverages or nutrition bars).There was concern when DSHEA was being drafted that somehow drugs would turn up as dietary supplements given that many drugs evolved from plants. For example, the ephedra plant brought the drug ephedrine and the Yew tree the cancer treating drug paclitaxel. The last sentence should be replaced by: Therefore, the drug inclusionary and exclusionary clause in DSHEA in subparagraph (2)(B) above protects the interests of the pharmaceutical industry.

In short, a dietary supplement can remain as such even if the substance is subsequently approved as a drug. However, a new drug or an authorized investigational new drug cannot be subsequently marketed as a dietary supplement.

After DSHEA became law, what became a very successful dietary supplement entered the market. "Cholestin," promoted to address high cholesterol, contained a botanically derived ingredient, red yeast rice that naturally contains the same cholesterol-lowering new drug ingredient as lovastatin. FDA argued that the red yeast rice was specially manufactured to contain high levels of the drug ingredient and that the ingredient was not marketed as a food or dietary supplement prior to new drug approval.

The key matter in this case was the designation of lovastatin or red yeast rice as the relevant "article" under the dietary supplement exclusionary provision quoted above. While the FDA considered lovastatin to be the relevant "article," the Cholestin manufacturer argued for red yeast rice to be the relevant "article."

The court found for FDA, concluding that the relevant "article" was lovastatin. Although Red Yeast Rice had been marketed as a constituent of traditional Asian foods prior to the approval of lovastatin, the court determined that the previously marketed products must have been promoted for their lovastatin content, or the lovastatin must have been increased or optimized as a constituent and that these situations had not occurred. Cholestin was therefore found to be a drug, not a dietary supplement, because it contained lovastatin, an article that was not marketed as a dietary supplement or food before it was approved as a drug. (Pharmanex Inc. v. Shalala, 35 F. Supp. 2d 1341 (D. Utah 1999), rev'd and remanded, 221 F.3d 1151 (10th Cir. 2000))

9.3 WHAT ARE NEW DIETARY INGREDIENTS?

DSHEA recognized that the dietary supplement industry was based both in traditional products such as vitamins and botanicals long available as supplements to the diet, and innovators in products to promote good health. DSHEA's new dietary ingredient provision addresses new ingredients including those predicated on ingredients that are food or used in food. First, what is a new dietary ingredient?

9.3.1 NEW DIETARY INGREDIENTS

For purposes of this section, the term "new dietary ingredient" means a dietary ingredient that was not marketed in the United States before October 15, 1994 and does not include any dietary ingredient which was marketed in the United States before October 15, 1994. From this definition comes two definitions, that of an old dietary ingredient as well as that of a new dietary ingredient. It is important to note that an old dietary ingredient must have been marketed "in the United States" prior to the trigger date, October 15, 1994 as a "dietary ingredient"—a category of food first defined by DSHEA.

New dietary ingredients can fall into two different classes. If a dietary ingredient is already present in the world's food supply in a chemically unaltered form, it can be freely included in dietary supplements without being deemed adulterated. If a new dietary ingredient does not fall in this first class, it needs to go through the notification process with evidence of its safety at least 75 days before commercialization. The legal language is provided below:

(1) The dietary supplement contains only dietary ingredients which have been present in the food supply as an article used for food in a form in which the food has not been chemically altered.
(2) There is a history of use or other evidence of safety establishing that the dietary ingredient when used under the conditions recommended or suggested in the labeling of the dietary supplement will reasonably be expected to be safe and, at least 75 days before being introduced or delivered for introduction into interstate commerce, the manufacturer, or distributor of the dietary ingredient or dietary supplement provides the Secretary with information, including any citation to published articles, which is the basis on which the manufacturer or distributor has concluded that a dietary supplement containing such dietary ingredient will reasonably be expected to be safe.

Under Section (1) above, the food supply is read to mean the world's food supply because more limiting language ("in the United States") was used to describe old dietary ingredients. Application of this definition to a country's food supply is limited to an ingredient's use in conventional foods as many foreign countries now have categories of "novel" ingredients and foods that are different than conventional foods.

FDA has written and implemented a regulation under which such notifications are made (21 CFR 190.6). If the applicable sections of the New Dietary Ingredient provision of the law are not met, then the dietary supplement containing such ingredient may be deemed to be adulterated and may be subject to an FDA enforcement action if and when it is marketed. This is a notification provision, not an approval provision. Once the notification is made, the product may be brought to market after 75 days.

In August 2016, FDA released a revised "Draft Guidance for Industry: Dietary Supplements: New Dietary Ingredient Notifications and Related Issues" in which FDA discusses the New Dietary Ingredient (NDI) provision in detail. This document is long (99 pages) and controversial. However, based on FDA's responses to NDI notifications, it appears that the Draft Guidance is being used by FDA. A listing of FDA responses to NDI notifications, as well as the publicly disclosable parts of the NDI notifications can be found in Regulations.gov.

In lieu of a notification, a petition may be filed proposing the issuance of an order prescribing the conditions under which an NDI is expected to be safe. The FDA is required to make a decision on such a petition within 180 days of its filing. To date, there have been no petitions filed under this provision.

9.4 HOW ARE DIETARY SUPPLEMENTS LABELED?

A dietary supplement label must include a statement of identity (e.g. "dietary supplement" or "herbal supplement"), an ingredient statement, nutrition information which is titled "Supplement Facts," and the name and place of business of the manufacturer, packer, or distributor, as well as an accurate statement of the quantity of contents, in terms of weight, measure, or numerical count.

FDA has published and updated a dietary supplement labeling guide that provides plain-language information regarding the labeling of dietary supplements with citations to the applicable 21 CFR regulation. This Guidance for Industry titled "A Dietary Supplement Labeling Guide" is broken into various sections: General dietary supplement labeling, identity statement, net quantity of contents, nutrition labeling, ingredient labeling, claims, premarket notification of new dietary ingredients, and other labeling information. The Guide also notes that the Tariff Act of 1930 requires that every article of foreign origin (or its container) imported into the United States conspicuously indicate the English name of the country of origin of the article.

One additional Federal labeling requirement is not noted in the Guide. The Dietary Supplement and Nonprescription Drug Consumer Protection Act which became law in 2006, and is discussed later in this chapter, requires that manufacturers and distributors include a domestic telephone number or domestic address on the label of their products through which they may receive a report of an adverse event. However, companies are not required to state that the information is for the purpose of reporting adverse events.

9.5 CLAIMS THAT MAY BE MADE FOR DIETARY SUPPLEMENTS

DSHEA created a category of claims unique to dietary supplements called "Statements of Nutritional Support," now commonly known as structure–function claims. The regulations for these claims can be found in 21 CFR 101.93. The FDA Federal Register notice accompanying the promulgation of 21 CFR 101.93 provides the most detailed guidance regarding the differences between structure–function claims and disease claims. It is important to note that making a structure–function claim is allowed for dietary supplements while making a disease claim would cause the product to be considered an unapproved drug. The final rule, "Regulations on Statements Made for Dietary Supplements Concerning the Effect of the Product on the Structure or Function of the Body," 65 Fed. Reg. 999. issued Jan. 6, 2000, contains details on this topic.

According to 21 USC 343(r)(6), a dietary supplement can make claims of a benefit-related addressing nutritional deficiency, structure–function, or general well-being as noted below:

(A) the statement claims a benefit related to a classical nutrient deficiency disease and discloses the prevalence of such disease in the United States, describes the role of a nutrient or dietary ingredient intended to affect the structure or function in humans, characterizes the documented mechanism by which a nutrient or dietary ingredient acts to maintain such structure or function, or describes general well-being from consumption of a nutrient or dietary ingredient.

The first, classical nutrient deficiency diseases, is one that is not often made because there are few classic nutrient deficiency diseases that are prevalent in the United States. Up-to-date information on such diseases is available from the Health and Medicine Division of the National Academy of Sciences. The last claim regarding general well-being (e.g. "contributes to good health") is used by some marketers.

Regarding the above claims, the law also requires that:

(B) the manufacturer of the dietary supplement has substantiation that such statement is truthful and not misleading.
It is important to note that the manufacturers are not obligated to provide this substantiation to FDA. In making a postmarketing notification to FDA regarding the claim, however, FDA requires that the company attest "that the notifying firm has substantiation that the statement is truthful and not misleading" (21 CFR 101.93(a)(3)). FDA has published guidance setting forth the Agency's views on the level of substantiation it believes to be required to substantiate such claims (Guidance for Industry: Substantiation for Dietary Supplement Claims Made Under Section 403(r) (6) of the Federal Food, Drug, and Cosmetic Act (Dec. 2008)). But the FDA has only rarely claimed in Warning Letters that a dietary supplement claim is not substantiated, thus challenging the recipient to provide substantiation.
Finally, the law requires the following to be clearly presented on the label regarding dietary supplement claims:

(C) the statement contains, prominently displayed and in boldface type, the following: "This statement has not been evaluated by the Food and Drug Administration. This product is not intended to diagnose, treat, cure, or prevent any disease."

In 21 CFR 101.93(b)–(e), the FDA sets forth how this disclaimer must be made. For example, the disclaimer must be set in a box in bold face type.

As mentioned earlier, a dietary supplement may make a claim that describes the role of a nutrient or dietary ingredient intended to affect the structure or function in humans, characterizes the documented mechanism by which a nutrient or dietary ingredient acts to maintain such structure or function. If the manufacturer of a dietary supplement proposes to make such a statement, the manufacturer must notify the FDA no later than 30 days after the first marketing of the product. If FDA objects to the claim, it will send a "Courtesy Letter" explaining why the claim is considered by FDA to be unlawful. This Courtesy Letter provides fair warning of FDA's position with regard to the claim. The Courtesy Letter, will go into the facility file of the relevant FDA District Office for inspection follow-up. There is a risk that FDA will initiate an official enforcement action in the future. However, note that FDA's failure to send a Courtesy Letter does not indicate "approval" or prevent FDA from challenging the claim in the future. FDA Courtesy Letters may be found on Regulations.gov.

A dietary supplement that states or implies "disease claims" may be subject to enforcement action as a drug and/or as a dietary supplement that is labeled with an unauthorized health claim. Distinguishing between lawful structure–function claims and disease claims is an area of great contention between marketers and FDA. In addition to 21 CFR 101.93, FDA has also published Guidance for Industry: Structure–Function Claims; Small Entity Compliance Guide (2002) which adds further information on this subject.

Where FDA disagrees strongly, a Warning Letter will likely be issued. Warning letters are available on FDA's web site, FDA.gov, and there are tools available to search for those letters that evaluate dietary supplement claims. These Warning Letters provide relatively "real-time" information regarding claims being made within this industry and challenged by FDA.

9.5.1 HEALTH CLAIMS

Health claims may also be made for dietary supplements meeting the requirements for these claims. Health claims characterize the relationship of a conventional food or dietary supplement to a disease- or health-related condition and are described in 21 CFR 101.14. FDA has promulgated regulations setting forth the specific requirements for the use of health claims, including the permissible language for each claim (21 CFR 101.70–101.83). An example of an approved health claim relevant to dietary supplements is: "Women who consume healthful diets with adequate folate throughout their childbearing years may reduce their risk of having a child with a birth defect of the brain or spinal cord. Sources of folate include fruits, vegetables, whole grain products, fortified cereals, and dietary supplements" (21 C.F.R. § 101.79).

FDA issues health claim regulations on its own or in response to a petition. A health claim may be authorized only if FDA determines that there is significant scientific agreement among experts qualified by scientific training and experience to evaluate such claims (based on the totality of scientific evidence) and that the claim is supported by appropriate evidence. FDA also may revise or revoke a health claim regulation. In October 2017, FDA made a proposal to revoke the health claim regarding the relationship between soy protein and coronary heart disease (21 CFR 101.82) because the totality of publicly available scientific evidence currently available did not support FDA's previous determination (U.S. Food and Drug Administration, 2017a, 2017b, 2017c, 2017d, 2017e, 2017f, 2017g, 2017h, 2017i, 2017j, 2017k, 2017l, 2017m, 2017n, n.d.).

The standard for substantiation of health claims is "significant scientific agreement." After a series of lawsuits over the rejection of petitions for health claims where the Courts ruled that there had to be another tier of claims below the regulatory standard to survive First Amendment scrutiny, FDA issued a Guidance for Industry: Evidence-Based Review System for the Scientific Evaluation of Health Claims (Jan. 2009). Under this guidance, FDA reviews petitions and has accepted certain "qualified" health claims for use in food and dietary supplement labeling. These are not codified in the Code of Federal Regulations but are found in a listing on FDA's website under Qualified Health Claims. An early and important example of a qualified claim is: "Supportive but not conclusive research shows that consumption of EPA and DHA omega-3 fatty acids may reduce the risk of coronary heart disease." These qualified health claims are set forth in Letters of Enforcement Discretion issued in response to submissions petitioning for these claims. FDA has also provided a guidance to industry with respect to submissions in support of qualified health claims entitled Guidance for Industry: FDA's Implementation of "Qualified Health Claims: Questions and Answers; Final Guidance (May 2006)".

9.5.2 NUTRIENT CONTENT CLAIMS

Nutrient content claims focus on specific nutrients through label, labeling, and advertising references. FDA regulates claims such as "good source of Vitamin E" or "low in fat" or "healthy." The applicable regulations appear in 21 CFR 101.13 and 21 CFR 101.54–101.69. These regulations have specific requirements to make "source" claims for nutrients or other ingredients. Antioxidant claims, often made for various dietary supplements, require special attention because the regulations limit "antioxidant" nutrient content claims to specific recommended daily intake (RDI) nutrients, for example, Vitamin C, at specific amounts, for example, greater than 10% of the RDI for the nutrient, which must also be named in the claim (21 CFR 101.54(g)). This regulation does not, however, bar an antioxidant activity claim which is a permitted structure–function claim, for example, curcumin provides antioxidant activity.

For claims about an ingredient that is not defined in the regulations or for which FDA has not established an RDI or DRV, dietary supplements may make a claim about a non-RDI/DRV nutrient that characterizes the percentage level of a dietary ingredient (e.g. "contains 200% of the lycopene in a tomato") (21 USC 343(r)(2)(F)). Moreover, claims that set forth the amount of a dietary ingredient are permitted, for example, "Contains 2 grams Omega-3 fatty acids (EPA/DHA) per serving" (21 CFR 101.13(i)(3) and 101.54(c)(1)).

One part of DSHEA that was a "must have" for the dietary supplement industry at the time the law was under consideration is a provision known as the published literature exemption. 21 USC 403B provides:

(a) *IN GENERAL—A publication, including an article, a chapter in a book, or an official abstract of a peer-reviewed scientific publication that appears in an article and was prepared by the author or the editors of the publication, which is reprinted in its entirety, shall not be defined as labeling when used in connection with the sale of a dietary supplement to consumers when it(1)is not false or misleading;(2)does not promote a particular manufacturer or brand of a dietary supplement;(3)is displayed or presented, or is displayed or presented with other such items on the same subject matter, so as to present a balanced view of the available scientific information on a dietary supplement;(4)if displayed in an establishment, is physically separate from the dietary supplements; and(5)does not have appended to it any information by sticker or any other method.*

(b) *APPLICATION—Subsection (a) shall not apply to or restrict a retailer or wholesaler of dietary supplements in any way whatsoever in the sale of books or other publications as a part of the business of such retailer or wholesaler.*

> (c) *BURDEN OF PROOF—In any proceeding brought under subsection (a), the burden of proof shall be on the United States to establish that an article or other such matter is false or misleading.*

This provision was important because of a series of enforcement actions where FDA asserted that displaying and selling books that misleadingly recommend an article of food as a remedy for various ailments in the same shop as the food constitutes misbranding in violation of 21 U.S.C. § 331 (a); that is, misleading written matter is accompanying the corresponding food.

Under the Third-party Literature provision, literature meeting the requirements of this section is not considered labeling and thus cannot be used by FDA as evidence of making a disease claim in labeling. Note, however, that FDA has taken the position that the use of published literature, including citations to such literature, especially on web sites regarding the product, can evidence the intent to market the product as a drug. FDA then asserts that the product is not generally recognized as safe and effective for the published literature referenced uses and, therefore, is a "new drug" under section 21 USC 321(p). New drugs may not be legally introduced or into interstate commerce without prior approval from FDA as required by, 21 USC 331(d) and 355(a). FDA uses this theory to make an end run around the published literature exemption.

9.6 MANUFACTURING AND DISTRIBUTING DIETARY SUPPLEMENTS

9.6.1 FACILITY REGISTRATION

Under the "Bioterrorism Act," as amended by the Food Safety Modernization Act (FSMA), owners, operators, or agents in charge of domestic or foreign facilities that manufacture, process, pack, or hold food, including dietary supplements for United States consumption, must register their facility with FDA (21 CFR Part 1 Subpart H). Retail establishments are exempt from registration. Food facilities that are required to register with FDA must renew their registrations every 2 years (between October 1 and December 31 of each even-numbered year).

The dietary supplement industry relies on botanicals and other ingredients that are grown, processed, or manufactured overseas. Foreign facilities that manufacture, process, pack, or hold food, including ingredients intended for use in dietary supplements, or dietary supplements, are required to register and renew unless food from that facility undergoes further processing (including packaging) by another foreign facility before the food is exported to the United States. If the subsequent foreign facility performs only minimal activity, however, such as putting on a label, both facilities are required to register (21 CFR 1.226(a)).

9.6.2 CURRENT GOOD MANUFACTURING PRACTICES FOR DIETARY SUPPLEMENTS

A dietary supplement is adulterated if it has been prepared, packed, or held under conditions that do not meet current good manufacturing practice (cGMP) regulations

(21 USC 342(g)(1)). FDA promulgated cGMP regulations for dietary supplements on June 25, 2007 (72 Fed. Reg. 34,751; 21 C.F.R. Part 111). Companies of all sizes are now required to be in compliance. The focus of the regulations is the requirement for each manufacturer to implement a system of production and process quality controls that covers all stages of manufacturing, packaging, labeling, and holding of the dietary supplement to ensure dietary supplements are manufactured, packaged, and labeled as specified in the master manufacturing record required for each product.

DSHEA provides that cGMP "regulations shall be modeled after current good manufacturing practice regulations for food and may not impose standards for which there is no current and generally available analytical methodology" (21 USC 342(g)(2)). At the time DSHEA became law (1994) and the cGMP regulations were promulgated (2007), the cGMP regulations covering food were those that had applied to food manufacturing for decades (21 CFR Part 110).

The cGMP regulations that FDA promulgated were nonetheless modeled more on the pharmaceutical model than on the food model. Extraordinarily, these regulations are longer and more detailed than those applicable to pharmaceuticals. Compare 21 CFR Part 111 to 21 CFR Part 211. In 21 CFR 111.75, FDA inspectors have imposed requirements that analytical testing be developed where none now exist. This issue is one that severely disrupts the businesses of herbal products manufacturers large and small, and only the future will tell whether FDA will challenge manufacturers who push back on this issue by invoking DSHEA's mandate against "standards for which there is no current and generally available analytical methodology." Note, FDA recognized the lack of available methodology in determining expiration dating would not be required in cGMPs for dietary supplements.

FDA's cGMP position on expiration or shelf-life dating is a trap for the unwary dietary supplement manufacture. FDA did not propose expiration dating for dietary supplements because, while "there are current and generally available methods to determine the expiration date of some dietary ingredients, for example, vitamin C, we are uncertain whether there are current and generally available methods to determine the expiration dating of other dietary ingredients, especially botanical dietary ingredients" (68 Fed. Reg. 12,157, 12,203-04 (Mar. 13, 2003)). In the final cGMP rule, FDA confirmed this decision. Nonetheless, as a practical matter, expiration dating is expected by retailers. FDA stated that "any expiration date that you place on a product label (including a 'best if used by' date) should be supported by data" (72 Fed. Reg. 34,855-56). This statement by FDA is not in the cGMP regulations, and importantly there is no regulatory requirement that such data be maintained and made available to FDA inspectors. The snare for the unwary exists because FDA investigators ask for this information and if it is provided, they will examine it, question it, and possibly deem it not supportive of the date placed on products. So, manufacturers providing expiration dating information to FDA should be aware of the risk.

FDA Warning Letters are published weekly by FDA on its website. The link is on the FDA home page. There are often Warning Letters to dietary supplement companies regarding cGMP violations and these provide a rich source of information regarding FDA's interpretation of the cGMP regulations and enforcement observations, which can also be found on FDA's web site. These letters are an important source of reminders and information for dietary supplement facility management and quality personnel.

9.6.3 FSMA AND DIETARY SUPPLEMENTS

Under the Food Safety Modernization Act, food facilities are required to develop Hazard Analysis and Risk-Based Preventive Controls programs to help ensure food safety. This requirement does not apply to a facility that manufactures, processes, packs, or holds dietary supplements that are in compliance with the cGMP requirements and adverse event reporting requirements for dietary supplements. However, FSMA-mandated requirements, 21 CFR 117, apply to the manufacture of the ingredients used in dietary supplements (e.g. dietary ingredients, food additives, GRAS substances, color additives, or other constituents that are used for their technical or functional effect or in the manufacture of supplements) because they are not subject to the dietary supplement cGMPs until they arrive at the manufacturer for use.

FSMA also requires importers of food, including ingredients used in the manufacture of dietary supplements, to implement foreign supplier verification programs to ensure that imported products meet US legal requirements. There is no statutory exception for dietary supplement manufacturers. However, FDA regulations implementing the foreign supplier verification program requirements include modified requirements for dietary supplement firms that are unable to comply with certain requirements (21 CFR 1.511). Similarly, dietary supplement manufacturers are not exempt from that part of FSMA, that for the first time, empowers FDA to order recalls if a company refuses to recall in situations where FDA has found a reasonable probability that the product will cause serious adverse health consequences or death to humans.

9.7 ADVERSE EVENT REPORTING FOR DIETARY SUPPLEMENTS

Adverse event reporting has long been the norm for new drugs and the requirement for reporting is built into the conditions of approval that accompany FDA new drug approvals. Adverse events reported to FDA are available under the Freedom of Information Act. Adverse events associated with food have traditionally been the domain of the Centers for Disease Control and Prevention, which monitors outbreaks of food poisoning, usually microbiological contamination-based and tracks down the source. FDA initiated adverse event monitoring for special nutritional products, including dietary supplements, in the early 1990s. This system received reports of adverse events associated with dietary supplements and a substantial number of reports concerned supplements containing ephedrine alkaloids, which became the starting point in the review of the safety of this substance and its eventual ban in 2004 (21 CFR 119.1).

In 2002, FDA announced that this system was limited and the information it provided was difficult to evaluate. FDA then developed a new system for tracking and analyzing adverse event reports involving foods, medical foods, cosmetics, and dietary supplements. In December

2016, FDA announced that adverse event data reported for these is now available on FDA's web site through the CFSAN Adverse Event Reporting System (CAERS).

9.7.1 DIETARY SUPPLEMENT AND NONPRESCRIPTION DRUG CONSUMER PROTECTION ACT

The ephedrine alkaloid experience was a teaching moment for the dietary supplement industry. First, the industry learned that health-related complaints about their products got the attention of FDA, state and local jurisdictions, the press, and the public. Second, the industry learned that the alleged lack of a direct causation link from a health-related complaint to the product was not a "defense" when the number of reports is substantial. Growing up in its regulated environment, the industry matured and came to understand that a mandatory system for adverse event reporting would be to its benefit. Industry engaged with both supporters and detractors in Congress in the pursuit of legislation mandating such a system.

The Dietary Supplement and Nonprescription Drug and Consumer Protection Act became law in December 2006. According to the new law, dietary supplement manufacturers, packers, or distributors must report serious adverse events that may be associated with the use of their dietary supplements to FDA. Serious adverse events (SAEs) include death, a life-threatening experience, inpatient hospitalization, a persistent or significant disability or incapacity, or a congenital anomaly or birth defect; or require, based on reasonable medical judgment, a medical or surgical intervention to prevent a serious outcome. These events must be reported to FDA within 15 business days of receipt by the company. The reports and records (telephone notes, emails, and correspondence) associated with these reports must be kept, along with all nonserious adverse events for 6 years.

Reports of adverse events associated with the use of dietary supplements, like those for drugs, are reports of association only and no cause-and-effect relationship needs to be present or found. As newcomers to this reporting process, companies are usually reluctant to report. They might not be aware that submission of the report by a company is not regarded as an admission that the dietary supplement involved caused or contributed to the adverse event being reported. The law also requires that manufacturers and distributors include a domestic telephone number or domestic address on the label of their products through which they may receive a report of an adverse event. However, the law does not require that there be a "call-out" on the label that telephone number or the address are for adverse event reporting. Under the law, adverse event report records must be made available to FDA inspectors upon request. FDA investigators usually make a request for these files as part of their opening requests to management. These files are reviewed to assure that the company has adequately investigated reports of adverse events and reported events that meet the definition of serious. In addition, FDA investigators will look at such reports to determine whether the company investigated the report in connection with their complaint investigation obligations under the cGMP regulations for dietary supplements (21 CFR 111.560).

FDA has not issued regulations under the Dietary Supplement and Nonprescription Drug and Consumer Protection Act, but has issued two guidance documents: Guidance for Industry: Questions and Answers Regarding Adverse Event Reporting and Recordkeeping for Dietary Supplements as Required by the Dietary Supplement and Nonprescription Drug Consumer Protection Act (Sept. 2013); and Guidance for Industry: Questions and Answers Regarding the Labeling of Dietary Supplements as Required by the Dietary Supplement and Nonprescription Drug Consumer Protection Act (Sept. 2009).

9.8 THE FEDERAL TRADE COMMISSION AND THE STATES, AND THE ADVERTISING AND MARKETING OF DIETARY SUPPLEMENTS

The Federal Trade Commission (FTC) "regulates" product advertising through guides, some regulations, but mostly through enforcement. The FTC and FDA work together under an agreement governing the division of responsibilities between the two agencies. FDA has the primary responsibility for regulating claims in labeling and the FTC regulates claims in advertising. Previously, claims found in advertising on television, radio, and print were regulated and enforced primarily by the FTC, whereas claims on labels and brochures were regulated by FDA. That has all changed now, with the heavy promotion of products like dietary supplements on the Internet and through promotional videos, social media, and social web sites. These platforms have been considering either advertising or labeling, or both. For this now dominant marketing method, both the FTC and FDA have taken enforcement actions.

In 1998, the FTC published a guide, "Dietary Supplements: An Advertising Guide for Industry." This guide has been cited by Federal courts as defining the substantiation requirement for claims made for dietary supplements. In the guide, the FTC reiterates the basic principles it applies to advertising for any product: (1) All advertising must be truthful and not misleading and (2) before disseminating an advertisement, advertisers must have adequate substantiation for all claims (the Prior Substantiation Doctrine). Consistent with its policy for health claims for other products such as foods and drugs, claims for dietary supplements will generally require substantiation with "competent and reliable scientific evidence." This is defined as: "Tests, analyses, research, studies, or other evidence based on the expertise of professionals in the relevant area that have been conducted and evaluated in an objective manner by persons qualified to do so, using procedures generally accepted in the profession to yield accurate and reliable results."

There is no fixed formula for the type and number of studies a manufacturer must have to substantiate its claims. In most cases, if the weight of the evidence does not support the claim, it should not be made, even if qualified. In substantiating a claim, all relevant research relating to the claimed benefit of their product should be considered, not simply the research that supports the effect, while disregarding research that does not.

In recent years, the FTC has been quite active in taking enforcement action against dietary supplement manufacturers and retailers making false, misleading, or unsubstantiated claims about their products. Some dietary supplement marketers have been required to pay a consumer redress of over $20 million or more, due to false or misleading advertising claims made in advertising materials challenged by the FTC. According to the FTC, "redress" means sellers must provide refunds to consumers the value corresponding to the sales, not the net profits. It's all about the money and the FTC is adept at finding it. In a number of cases, the FTC has required the sale of company owners' homes, cars, and boats as part of the resolution of the case.

9.8.1 THE FTC AND "THE SELF-REGULATORY PROCESS"

The National Advertising Division of the Council of Better Business Bureaus has a process whereby complaints against advertising may be made by competitors and indeed, almost anyone. The collection of programs is denominated the Advertising Self-Regulatory Council. Much of the Council staff are former or retired FTC staff with substantial experience in advertising review. The product advertising evaluated by this process covers the entire range of products marketed to consumers. Because of the large number of private bar, private plaintiff lawsuits that result after the publication of decisions that come out of this self-regulatory process, many companies now settle their competitive disputes privately to avoid the highly likely class action lawsuits that follow.

9.8.2 STATE ENFORCEMENT AND THE PRIVATE BAR

In addition to the FDA and the FTC, state attorneys general and the private bar are also active in enforcing against false or misleading dietary supplement claims, and such actions often result in the payment of monetary damages. For example, in California, the Country Attorney's offices have for some time used California state consumer protection laws to pursue cases against dietary supplement marketers large and small. Where the case is large enough, the State Attorney General will also become involved, and sometimes even the Federal Trade Commission. State Attorneys General in New York and Oregon have also actively pursued companies marketing dietary supplements. The New York Attorney General sent letters and threatened action against various companies that it claimed could not substantiate the label claims for their botanical products. The threatened actions were predicated on the failure of products to pass DNA testing of botanical products commissioned by the State. In Oregon, the Attorney General pursued similar cases against major industry players regarding the sale of supplements with allegedly unsafe or unlawful ingredients.

The private bar and their private plaintiffs also pursue cases against dietary supplement companies. Some individuals and small firms make their livings by writing to companies detailing the false and misleading nature of the claims they make for products. They then settle with the company for small amounts that add up to large amounts because they send out lots of these demand letters. Other private bar actions include the filing of class action lawsuits, usually against larger companies and retailers with regard to products generating substantial sales.

REFERENCES

National Institutes of Health. (n.d.). *Office of Dietary Supplements. Dietary Supplement Health and Education Act of 1994. Public Law 103-417. 103rd Congress*. Retrieved from https://ods.od.nih.gov/About/DSHEA_Wording.aspx.

Public Law. (1994). *Public Law 108-417*, OCT. 25.

U.S. Food and Drug Administration. (n.d.). *Draft guidance for industry: Dietary supplements—new dietary ingredient notifications and related issues*. Retrieved from https://www.fda.gov/food/guidanceregulation/guidancedocumentsregulatoryinformation/dietarysupplements/ucm257563.htm.

U.S. Food and Drug Administration. (2017a). *FDA Food Safety Modernization Act (FSMA)*. Retrieved from https://www.fda.gov/food/guidanceregulation/fsma/default.htm.

U.S. Food and Drug Administration. (2017b). *Draft guidance for industry: Substantiation for dietary supplement claims made under Section 403 (r) (6) of the Federal Food, Drug and Cosmetic Act*. Retrieved from https://www.fda.gov/food/guidanceregulation/guidancedocumentsregulatoryinformation/dietarysupplements/ucm073200.htm.

U.S. Food and Drug Administration. (2017c). *Guidance for industry: FDA's implementation of "qualified health claims": Questions and answers; final guidance*. Retrieved from https://www.fda.gov/food/guidanceregulation/guidancedocumentsregulatoryinformation/ucm053843.htm.

U.S. Food and Drug Administration. (2017d). *21 CFR 101.93: Certain types of statements for dietary supplements*. Retrieved from CFR-Code of Federal Regulations Title 21, https://www.accessdata.fda.gov/scripts/cdrh/cfdocs/cfcfr/CFRSearch.cfm?fr=101.93.

U.S. Food and Drug Administration. (2017e). *21 CFR 101.14: Health claims: general requirements*. Retrieved from CFR-Code of Federal Regulations Title 21, https://www.accessdata.fda.gov/scripts/cdrh/cfdocs/cfcfr/cfrsearch.cfm?fr=101.14.

U.S. Food and Drug Administration. (2017f). *21 CFR 101.79: Health claims: folate and neural tube defects*. Retrieved from CFR-Code of Federal Regulations Title 21, https://www.accessdata.fda.gov/scripts/cdrh/cfdocs/cfcfr/cfrsearch.cfm?fr=101.79.

U.S. Food and Drug Administration. (2017g). *21 CFR 101.54: Nutrient content claims for "good source," "high," "more,"* and "*high potency*." Retrieved from CFR-Code of Federal Regulations Title 21, https://www.accessdata.fda.gov/scripts/cdrh/cfdocs/cfcfr/cfrsearch.cfm?fr=101.54.

U.S. Food and Drug Administration. (2017h). *21 CFR 1.226: Who does not have to register under this subpart?* Retrieved from CFR-Code of Federal Regulations Title 21, https://www.accessdata.fda.gov/scripts/cdrh/cfdocs/cfcfr/cfrsearch.cfm?fr=1.226.

U.S. Food and Drug Administration. (2017i). *21 CFR Part 110 : Current good manufacturing practices*. Retrieved from CFR-Code of Federal Regulations Title 21, https://www.fda.gov/food/guidanceregulation/cgmp/default.htm.

U.S. Food and Drug Administration. (2017j). *21 CFR Part 111: Current good manufacturing practice in manufacturing, packaging, labeling, or holding operations for dietary supplements*. Retrieved from CFR-Code of Federal Regulations Title 21, https://www.accessdata.fda.gov/scripts/cdrh/cfdocs/cfcfr/cfrsearch.cfm?cfrpart=111.

U.S. Food and Drug Administration. (2017k). *21 CFR Part 111: Current good manufacturing practice, hazard analysis, and risk-based preventive controls for human food*. Retrieved from CFR-Code of Federal Regulations Title 21, https://www.accessdata.fda.gov/scripts/cdrh/cfdocs/cfcfr/cfrsearch.cfm?cfrpart=117&showfr=1&subpartnode=21:2.0.1.1.16.7.

U.S. Food and Drug Administration. (2017l). *21 CFR Part 1.511: What FSVP must I have if I a importing a food subject to certain requirements in the dietary supplement current good manufacturing practice regulation?* Retrieved from CFR-Code of Federal Regulations Title 21, https://www.accessdata.fda.gov/scripts/cdrh/cfdocs/cfcfr/cfrsearch.cfm?fr=1.511.

U.S. Food and Drug Administration. (2017m). *21 CFR Part 119.1: Dietary supplements containing ephedrine alkaloids*. Retrieved from CFR-Code of Federal Regulations Title 21, https://www.accessdata.fda.gov/scripts/cdrh/cfdocs/cfcfr/cfrsearch.cfm?fr=119.1.

U.S. Food and Drug Administration. (2017n). *21 CFR Part 111.560: What requirements apply to the review and investigation of a product complaint?* Retrieved from CFR-Code of Federal Regulations Title 21, https://www.accessdata.fda.gov/scripts/cdrh/cfdocs/cfcfr/cfrsearch.cfm?fr=111.560.

FURTHER READINGS

Government Printing Office. (2018). *21 CFR 101.82: Health claims: Soy protein and risk of coronary heart disease (CHD)*. Retrieved from Electronic Code of Federal Regulations Title 21, https://www.ecfr.gov/cgi-bin/text-idx?SID=57afc460a43ec19a1dd61701b90828a6&mc=true&node=se21.2.101_182&rgn=div8.

LLI/Legal Information Institute. (n.d.a). *U.S. Code Title 21. 21 USC 321(ff)—Definitions; generally*. Retrieved from https://www.law.cornell.edu/uscode/text/21/321.

LLI/Legal Information Institute. (n.d.b). *U.S. Code Title 21. 21 USC 331—Prohibited acts*. Retrieved from https://www.law.cornell.edu/uscode/text/21/331.

LLI/Legal Information Institute. (n.d.c.). *U.S. Code Title 21. 21 USC 343: Adulterated food*. Retrieved from https://www.law.cornell.edu/uscode/text/21/342.

LLI/Legal Information Institute. (n.d.d). *U.S. Code Title 21. 21 USC 343: Misbranded food*. https://www.law.cornell.edu/uscode/text/21/343.

CHAPTER

10

Cosmetics

Simone E. Turnbull
Sanofi, Chattanooga, TN, United States

"These products must be evaluated by FDA as drugs before the companies can make claims about changing the skin or treating disease."

– Linda M. Katz

An Overview of FDA Regulated Products. http://dx.doi.org/10.1016/B978-0-12-811155-0.00010-7

CHAPTER OBJECTIVES

After reading this chapter, the reader will be able to:

- identify what constitutes a cosmetic, drug, and a cosmetic-drug,
- identify what causes a cosmetic to be adulterated or misbranded,
- describe the regulatory pathway for placing a cosmetic product on the US market,
- identify the labeling requirements for cosmetics,
- discern the difference between cosmetic and unapproved drug claims, and
- compare/contrast the US and international approaches to cosmetic regulation.

10.1 IS IT A COSMETIC, A DRUG, OR A COSMETIC-DRUG?

Cosmetics are a subset of personal care products commonly used by consumers as part of their daily activities of hygiene and beautification. Like drugs, they are defined by their intended use. If the intent is to cleanse and beautify the skin, hair, or nail from an appearance perspective, then that product is a cosmetic. However, if the intent of a product is to do more than beautification, that is, affect the human body in a therapeutic manner, then that product is considered as a drug and must meet the requirements for drugs. Cosmetics are regulated by the Center for Food Safety and Applied Nutrition (CFSAN) within the US Food and Drug Administration (FDA). CFSAN also regulates food and dietary supplement products. Drugs are regulated by the Center for Drug Evaluation and Research (CDER), also within the FDA.

Cosmetic-drugs are most commonly over-the-counter (OTC) drugs that fall within the scope of definitions of both cosmetic and drug. Therefore, they must meet the regulatory requirements for both. For example, an antidandruff shampoo is a cosmetic because its intended use is to cleanse the hair. However, it is also a drug because it is intended to control and help prevent the recurrence of dandruff symptoms, for example, flaking and itching. Other examples of cosmetic-drugs include toothpastes that contain fluoride, deodorants that are also antiperspirants, and moisturizers and makeup products marketed with sun-protection claims. It is important to distinguish cosmetic-drugs from "cosmeceuticals". The latter is an industry term meant to describe a cosmetic with therapeutic benefits. However, any cosmetic that is marketed as having drug properties is considered to be a drug and must meet the requirements for drugs (US Food and Drug Administration, 2015a). Both CFSAN and CDER may bring regulatory enforcement action against the manufacturers or distributors of cosmetic products marketed with drug claims.

10.2 HOW ARE COSMETICS DEFINED?

In the late nineteenth century, cosmetics were crude in-home preparations that were shared socially among women as part of America's beauty culture (Peiss, 1998). In the early twentieth century, this shifted to a mass-market industry. After a series of publicized incidents of consumer injuries and poisonings from lead, mercury, and other harmful ingredients, President Franklin D. Roosevelt signed the (Copeland–Lea) Federal Food, Drug and Cosmetic Act (FD&C Act or "the Act") into law in 1938 (Newburger, 2009; Termini & Tressler, 2008). Cosmetics were introduced into the Act, and along with food and drugs they were now subject to the regulatory authority of the FDA. As defined in the Act, cosmetics are

> *articles intended to be rubbed, poured, sprinkled, or sprayed on, introduced into, or otherwise applied to the human body or any part thereof for cleansing, beautifying, promoting attractiveness, or altering the appearance (U.S.C., 2010b).*

Among the products included in this definition are skin moisturizers, perfumes, lipsticks, fingernail polishes, eye and facial makeup products, cleansing shampoos, permanent waves, hair colors, and deodorants, as well as any substance intended for use as a component of a cosmetic product. Not included in this definition is soap, which is regulated by the Consumer Product Safety Commission (CPSC) (US Food and Drug Administration, 2016b).

In contrast to the definition of cosmetics, the FD&C Act defined drugs by their intended use as

> *articles intended for use in the diagnosis, cure, mitigation, treatment, or prevention of disease and articles (other than food) intended to affect the structure or any function of the body of man or other animals (U.S.C., 2010a).*

According to these definitions, the legal difference between a cosmetic and a drug is determined by the intended use of a product. In the FDA guidance "Is It a Cosmetic, a Drug, or Both? (Or Is It Soap?)", the intended use of a product can be established in a number of ways including claims, consumer perception, and known therapeutic uses of ingredients.

10.2.1 CLAIMS

Claims refer to information on product labeling, in advertising, on the Internet, or in other promotional materials. Certain claims may cause a product to be considered a drug, even if the product is marketed as if it were a cosmetic. If the intended use of a product is to treat or prevent disease or otherwise affect the structure or function of the human body, it would be classified as a drug. Some examples are products that claim to restore hair growth, reduce cellulite, treat varicose veins, increase or decrease the production of melanin (pigment) in the skin, or regenerate cells.

10.2.2 CONSUMER PERCEPTION

Consumer perception, established through a product's reputation, may reflect that the product does more than just beautification. In this case, the intended use may reach beyond what is allowed under cosmetic classification. Determining consumer perception may require asking why the consumer is buying the product and what the consumer expects it to do.

10.2.3 INGREDIENTS

If an ingredient of a product is known to be a drug or believed to have a therapeutic use, the product would then be considered a drug. An example is fluoride in toothpaste. In this case, a toothpaste containing fluoride would be classified as an OTC drug while one without fluoride or any other drug ingredient would be classified as a cosmetic (US Food and Drug Administration, 2016d).

10.3 HOW ARE COSMETICS REGULATED?

As the primary law governing all FDA-regulated products including cosmetics, the FD&C Act prohibits the marketing of adulterated or misbranded products in interstate commerce. The term adulterated refers to violations involving product composition. A cosmetic shall be deemed *adulterated*:

(a) If it bears or contains any poisonous or deleterious substance which may render it injurious to users under the conditions of use prescribed in the labeling thereof, or under such conditions of use as are customary or usual, except that this provision shall not apply to coal-tar hair dye
(b) If it consists in whole or in part of any filthy, putrid, or decomposed substance
(c) If it has been prepared, packed, or held under insanitary conditions whereby it may have become contaminated with filth, or whereby it may have been rendered injurious to health
(d) If its container is composed, in whole or in part, of any poisonous or deleterious substance which may render the contents injurious to health
(e) If it is not a hair dye and it is, or it bears or contains, a color additive which is unsafe (U.S.C., 2016b)

In contrast, misbranded refers to violations involving improperly labeled or deceptively packaged products. A cosmetic shall be deemed *misbranded*:

(a) *If its labeling is false or misleading in any particular*
(b) *If in package form unless it bears a label containing (1) the name and place of business of the manufacturer, packer, or distributor and (2) an accurate statement of the quantity of the contents in terms of weight, measure, or numerical count*
(c) *If any word, statement, or other information required by or under authority of this chapter to appear on the label or labeling is not prominently placed thereon with such conspicuousness (as compared with other words, statements, designs, or devices, in the labeling) and in such terms as to render it likely to be read and understood by the ordinary individual under customary conditions of purchase and use*
(d) *If its container is so made, formed, or filled as to be misleading*
(e) *If it is a color additive, unless its packaging and labeling are in conformity with such packaging and labeling requirements, applicable to such color additive*
(f) *If its packaging or labeling is in violation of an applicable regulation (U.S.C., 2016c)*

Other laws also affect the regulation of cosmetic and other FDA-regulated products. The Fair Packaging and Labeling Act (FPLA) of 1967 prevents unfair and deceptive packaging and labeling practices of consumer products. Under the FPLA, consumers should be able to obtain accurate information about the quantity of the contents contained within a package and should be able to compare values across products (U.S.C., 2016a). Enforcement of the FPLA falls under the jurisdiction of the Federal Trade Commission (FTC), whose dual mission is to protect consumers and promote competition (Federal Trade Commission, 2013). Failure to comply with the FPLA can cause a product to be misbranded and can result in enforcement action by either or both agencies.

10.3.1 PRODUCTS AND INGREDIENTS

The United States has a permissive regulatory system for cosmetics. With the exception of color additives, cosmetic products and their ingredients require neither notification to the FDA nor an approval by the FDA prior to being placed on the market. A manufacturer may use any ingredient in the formulation of a cosmetic, provided that

- the ingredient and the finished cosmetic are safe under labeled or customary conditions of use
- the product is properly labeled, and
- the use of the ingredient does not otherwise cause the cosmetic to be adulterated or misbranded under the laws that FDA enforces (US Food and Drug Administration, 2016b).

There are, however, specific FDA regulations that prohibit or restrict the use of the certain ingredients in cosmetics that can be found in Title 21, Part 700 of the Code of Federal Regulations (CFR).

10.3.1.1 Color additives

Color additives, with the exception of coal-tar hair dyes, are permitted in cosmetics only if the FDA has approved them for the intended use. Additionally, some color additives may be used only if they are from batches that the FDA has tested and certified. If a cosmetic product contains a color additive, it must meet FDA requirements for approval, certification, identity, and specification, and use and restrictions. Failure to meet US color additive requirements causes a cosmetic to be adulterated.

Color additives are divided according to two main categories: (1) color additives subject to batch certification, which are usually synthetically derived primarily from petroleum and are sometimes known as "coal-tar dyes" or "synthetic-organic" colors and (2) color additives exempt from batch certification, obtained primarily from natural sources like minerals, plants, or animals (US Food and Drug Administration, 2016a). Examples include annatto, caramel, and carmine. All color additives approved for use in cosmetics that are exempt from and subject to certification are identified in 21 CFR Part 73, Subpart C, and 21 CFR Part 74, Subpart C, respectively. The regulations also identify two other categories of color additives called straight colors and lakes. Straight colors refer to color additives that have not been mixed or chemically reacted with other substances. They are listed in 21 CFR Parts 73, 74, and 81 (CFR, 2016a). Lakes refer to colors that are formed from straight colors through adsorption, coprecipitation, or chemical combination. Lakes do not include any combinations of ingredients made by a simple mixing process (CFR, 2016b).

10.3.1.2 What about coal-tar hair dyes?

According to the FD&C Act, coal-tar hair dyes, unlike other color additives, are permitted on the market without FDA approval if the products provide adequate directions for consumers to conduct a skin test before they dye their hair. In addition, the product label must contain the following caution statement:

> *Caution—This product contains ingredients which may cause skin irritation on certain individuals and a preliminary test according to accompanying directions should first be made. This product must not be used for dyeing the eyelashes or eyebrows; to do so may cause blindness (FD&C Act, 601(a)).*

10.3.2 REGISTRATION AND MANUFACTURING REQUIREMENTS

In the United States, cosmetic registration is a voluntary system, which is in direct contrast with drug requirements where drug firms *must* register their establishments and list their drug products with FDA (U.S.C., 2010c). Manufacturers and distributors of cosmetic products have the option to register their establishments and cosmetic product ingredient statements (CPIS) with the FDA via the Voluntary Cosmetic Registration Program (VCRP).

Also, unlike drugs, cosmetics do not need to adhere to the current good manufacturing practices (cGMPs) requirements. Although not required, the FDA has provided guidelines for cosmetic GMPs in the form of an inspection checklist, which can be found on the FDA's website.

10.3.3 ADVERSE EVENT REPORTING OF COSMETICS

Currently there is no mandatory regulatory requirement to report adverse events (serious or nonserious) associated with the use of cosmetic products to the FDA. This is in contrast to the requirements for other regulated products like drugs and dietary supplements for which adverse event reporting is required. For cosmetics, the reporting is voluntary in that anyone, including consumers or healthcare professionals, can voluntarily report a complaint or adverse event (such as an illness, allergic reaction, rash, irritation, scarring, or hair loss) related to the use of a cosmetic product to the FDA.

10.3.4 FDA RECALL AUTHORITY FOR COSMETICS

Under the FD&C Act, the FDA has no authority to order the recall of a cosmetic product; they can only request that a cosmetic company recalls its product. This is in contrast to drug regulations whereby recalls may be conducted through a company's own initiative, at the request of the FDA, or a mandate by the FDA under their statutory authority. However, although the FDA can only request a cosmetic product recall, they do monitor the status of the recall once initiated. Additionally, they assign the risk classification level (Class I, II, or III, from most to least risk) posed by the recalled product (US Food and Drug Administration, 2016c).

It is the responsibility of the cosmetic company to notify their customers of a recall and to provide the appropriate FDA district office with periodic status updates. The cosmetic company is also responsible for monitoring the effectiveness of the recall and for deciding how the recalled product should be managed, including whether the product should to be destroyed or be brought into compliance.

10.3.5 COSMETIC LABELING AND LABEL CLAIMS

The FD&C Act defines labeling as all labels and other written, printed, or graphic matter on or accompanying a product. Cosmetic labeling should be truthful and not misleading. Since the FDA does not preapprove cosmetic labeling, it is the responsibility of the manufacturer and/or the distributor to ensure that their products are labeled appropriately. Failure to comply with the labeling requirements of the FD&C Act may cause a product to be misbranded.

Under the Act, cosmetics are required to bear certain information on the principal display panel (PDP) defined as part of the product label that is most likely to be displayed or examined by consumers under customary conditions of sale. Any panels other than the PDP that bear product information visible to consumers are called information panels. Required on the PDP are

- the statement of identity indicating the nature and use of the product and
- an accurate statement of the net quantity of contents, in terms of weight, measure, or numerical count.

Required on the information panels are

- the name and place of business: this may be the manufacturer, packer, or distributor;
- The distributor statement, if the name and place of business are not those of the manufacturer;
- any material facts, for example, directions for safe use, if a product could be unsafe if used incorrectly;
- any warning and caution statements, which must be prominent and conspicuous; and
- the ingredient listing, which must appear in descending order of predominance (CFR, 2016d).

Cosmetic-drugs must comply with the regulations for both OTC drug and cosmetic ingredient labeling. The active drug ingredient(s) must appear according to the OTC drug labeling requirements, and the inactive cosmetic ingredients must appear separately, in decreasing order of predominance (CFR, 2016c). Further labeling requirements for OTC drugs are discussed in Chapter 3.

10.3.6 DEMONSTRATING SAFETY AND EFFICACY

Under the US regulatory system for cosmetics, companies that market cosmetic products bear the responsibility of ensuring that their products are safe for consumers under labeled or customary conditions of use. Cosmetic companies that do not have adequate substantiation of safety on file are required to display the 21 CFR 740.10(a) warning statement on the PDP of their products, "Warning—The safety of this product has not been determined."

Cosmetic companies are also not required to provide proof of efficacy when placing their cosmetic product on the market. As a result, a cosmetic product can neither be labeled nor advertised as having FDA approval. This applies even if the manufacturing establishment is registered or the cosmetic product is on file with the VCRP. Similarly, cosmetic companies are not permitted to market their cosmetic product with therapeutic claims. If a cosmetic product is marketed as having drug properties, then the FDA must approve it as a drug. According to the FDA, some of the common unapproved drug claims made on products marketed as cosmetics are those for acne treatment, cellulite and stretch mark reduction, wrinkle removal, dandruff treatment, hair restoration, and eyelash growth (US Food and Drug Administration, 2015b).

The FDA has the authority to take regulatory action against companies marketing their cosmetic products with unapproved drug claims. As an initial level of enforcement action, either CDER or CFSAN can issue a warning letter to companies citing the unapproved drug claims. Upon receipt of a warning letter, a cosmetic company has 15 working days to respond to the FDA with the specific steps they have taken to correct the stated violations. A warning letter should be taken seriously. Failure to address the comments stated in a warning letter can escalate into more serious enforcement actions from the FDA, or even court actions through the Department of Justice (US Food and Drug Administration, 2015b).

10.4 INTERNATIONAL APPROACHES TO COSMETICS REGULATION

In order to provide a global regulatory perspective on cosmetics, it may be informative to compare the regulatory requirements for cosmetics and cosmetic-drugs in the United States with that of selected countries; in particular, Europe, Canada, and Japan. These countries are instructive examples because their regulations for cosmetics and cosmetic-drugs are mature and they, along with the United States, are voluntary members of the International Cooperation on Cosmetics Regulation (ICCR), which is comprised of regulatory authorities from each member country working to protect the consumer and minimize international trade barriers (Health Canada, 2014).

10.4.1 EUROPE

Similar to the United States, both cosmetics and drugs are defined by their intended use in the European Union (EU). The European regulations define cosmetics as "any substance or mixture intended for use on the body for purposes of cleansing, beautifying or perfuming" (Regulation, 2009). In contrast, drugs or medicinal products are defined as "any substance or combination of substances presented as having properties for treating or preventing disease in human beings" (Directive, 2001). Because OTC drugs do not exist in the EU, products like sunscreens and toothpastes are regulated there as cosmetics. But the EU regulations for cosmetics include requirements that are currently lacking in the United States, including the requirement for mandatory cosmetic product notification and communication of serious undesirable effects, akin to serious adverse event reporting in the United States.

10.4.1.1 Notification of cosmetic products

Cosmetic notification requirements in the EU resemble drug listing requirements in the United States, but they have some additional elements. European regulations require a "responsible person," a European person or business entity designated by the cosmetic manufacturer or distributor, to submit establishment and product listing information through the centralized Cosmetic Products Notification Portal (CPNP). This online system was specifically created for the implementation of the Cosmetic Regulation (EC) No 1223/2009 and provides cosmetic product information to the Competent Authorities of all member states for purposes of market surveillance, analysis, and consumer information (Regulation, 2009). In the EU, a regulation is a legal act that becomes immediately enforceable as law in all member states.

The regulation also requires companies to create a Cosmetic Product Safety Report (CPSR), which is divided into two main parts. Part A contains information related to the product composition, microbiological quality, and physical/chemical characteristics, and stability. Part B is the safety assessment conducted by a qualified EU Safety Assessor. The CPSR forms the foundation of the Product Information File (PIF). From this, one can easily derive or draw conclusions regarding the safety of a cosmetic product.

The PIF contains the quality, efficacy, and safety information for cosmetics and cosmetic ingredients manufactured in or imported into the EU. The responsible person is required to keep the PIF on file for 10 years from the date the last batch of the cosmetic product is placed on the market. In the event of a serious undesirable effect, he or she must make the PIF readily available to the competent authority of the Member State in which the PIF is kept.

10.4.1.2 Communication of serious undesirable effects

In Europe, it is mandatory to report serious undesirable effects (SUEs) in cosmetic products, unlike the United States where adverse event reporting is only required for drugs, biologics, medical devices, and dietary supplements. An SUE is defined as any effect that results in temporary or permanent functional incapacity, hospitalization, or death. The report must include the nature of the effect, the affected product, and any corrective action taken.

10.4.2 CANADA

Similar to the United States and the EU, Canada views a cosmetic as any substance used to clean, improve, or alter the appearance of skin, hair, nails, or teeth. Drugs are substances indicated for the diagnosis and treatment of a disease. Canadian regulations also acknowledge cosmetic-drugs, which are defined as "a subset of personal care products, which are not easily distinguished as either a drug or cosmetic" (Health Canada, 2008). Examples of cosmetic-drugs include acne medications, antiperspirants, antidandruff shampoos, sunburn protectants, and fluoride-containing anticaries products. Canada also recognizes another category of product called natural health products (NHPs). They are considered as a subset of drugs, which treat disease or modify bodily functions in a manner that maintains or promotes health. Depending on the active ingredient used, some cosmetic-drugs may be regulated as drugs or as NHPs. For example, a sunscreen containing either titanium dioxide or zinc oxide as an active ingredient is regulated as an NHP, whereas a sunscreen containing oxybenzone as an active ingredient is regulated as a drug product.

In Canada, Health Canada must be notified within 10 days after placing a cosmetic product on the market (Regulations, 2013). The required content of the notification is similar to that in the EU. Failure to notify a cosmetic product may result in that product being refused entry into Canada or removed from sale.

10.4.3 JAPAN

As in the other ICCR member countries, a product is deemed a cosmetic or a drug based on the intended use in Japan. Cosmetics are defined as

> *articles with mild action on the human body, which are intended to be applied to the human body through rubbing, sprinkling or other method, aiming to clean, beautify and increase the attractiveness, alter the appearance or to keep the skin or hair in good condition (Rannou, 2015).*

However, drugs are defined as

> *substances that are intended for use in the diagnosis, treatment, or prevention of disease in humans or animals, and substances that are intended to affect the structure or function of the body of humans or animals (Japan Pharmaceutical Manufacturers Association, 2015).*

Excluded from this definition are quasi-drugs, cosmetics, or regenerative medicine products. A *quasi-drug* is defined as an item for the purpose of

(1) preventing nausea and other discomfort
(2) preventing heat rash, soreness, etc.
(3) encouraging hair growth or removing hair, or
(4) exterminating and preventing mice, flies, mosquitoes, fleas, etc. (Rannou, 2015).

Examples of quasi-drugs include deodorants, depilatories, hair growth treatments, hair dyes, and agents for permanent hair waving and straightening. Medicated cosmetics are also quasi-drugs and they include whitening, acne prevention, and antiaging products, and products that disinfect the skin and mouth. Manufacturers and distributors of quasi-drugs are required to obtain premarket approval from the Japanese Ministry of Health, Labour and Welfare (MHLW) prior to placing their quasi-drug on the market, whereas manufacturers and distributors of cosmetic products are only required to notify the MHLW about the marketing of their cosmetic product prior to importation into Japan (Japan External Trade Organization (JETRO), 2011).

10.5 FUTURE REGULATORY OUTLOOK FOR COSMETICS IN THE UNITED STATES

In the United States, cosmetics enjoy a more permissive regulatory system than drugs, requiring neither notification nor proof of efficacy or safety prior to being placed on the market. They are not required to follow GMPs for manufacturing or to report adverse events associated with their uses. The regulatory requirements in the United States are, in fact, more lenient than those of other advanced economies.

With more potent cosmetic products entering the US market along with increasing reports of adverse events, several new legislations have been proposed over the past few years. These bills contain more stringent requirements, with some mirroring similar measures in place in other constituencies and others that go a step further. For example, these bills include provisions such as mandatory cosmetic product and manufacturing establishment registration, user fees, GMPs, and adverse event reporting (H.R.1385 2013–2014; S.1014, 2015–2016). Hence, the US framework for cosmetics may undergo significant changes over the next few years.

REFERENCES

CFR. (2016a). 21 C.F.R. § 70.3(j).

CFR. (2016b). 21 C.F.R. § 70.3(l).

CFR. (2016c). 21 C.F.R. § 201.66.

CFR. (2016d). 21 C.F.R. § 701.

Cosmetic Regulations. (2013). C.R.C. c. 869.

Directive. (2001). *2001/83/EC of the European Parliament and of the Council of 6 November 2001 on the community code relating to medicinal products for human use*. Consolidated version: 16/11/2012, OJ L 311, 28.11.2001 (p. 67). C.F.R.

Federal Trade Commission. (2013). *What we do. About the FTC*. Retrieved from http://www.ftc.gov/about-ftc/what-we-do.

Health Canada. (2008). *Guidance document: Classification of products at the cosmetic–drug interface*. Retrieved from http://www.hc-sc.gc.ca/cps-spc/alt_formats/hecs-sesc/pdf/pubs/indust/cosmet_drug_guide-drogue_ref/cosmet_drug_guide-drogue_ref-eng.pdf.

Health Canada. (2014). *International Cooperation on Cosmetics Regulation (ICCR). Information for industry and professionals—cosmetics and personal care products: Notices, guidelines, policies, reports, publications and international cooperation*. Retrieved from http://www.hc-sc.gc.ca/cps-spc/cosmet-person/indust/information/iccr/index-eng.php.

H.R.1385, 113th Congress (2013-2014).

Japan External Trade Organization (JETRO). (2011). *Guidebook for export to Japan 2011*. Retrieved from http://www.jetro.go.jp/mexico/mercadeo/1Ecosme.pdf.

Japan Pharmaceutical Manufacturers Association. (2015). *Pharmaceutical administration and regulations in Japan Regulatory Information Task Force International Affairs Committee*. Retrieved from http://www.jpma.or.jp/english/parj/pdf/2015.pdf.

Newburger, A. E. (2009). Cosmeceuticals: Myths and misconceptions. *Clinics in Dermatology*, 27(5), 446–452 http://dx.doi.org/10.1016/j.clindermatol.2009.05.008.

Peiss, K. L. (1998). *Hope in a jar: The making of America's beauty culture*. Philadelphia, PA: First University of Pennsylvania Press (2011 ed.).

Rannou, E. (2015). *Guidebook for exporting/importing cosmetics to Japan*. Retrieved from http://cdnsite.eu-japan.eu/sites/default/files/publications/docs/cosmetics-japan.pdf.

Regulation. (2009). *EC No. 1223/2009 of the European parliament and of the council of 30 November 2009 on cosmetic products (recast)*. OJ L 342, 22.12.2009 (p. 59–209). C.F.R.

S.1014, 114th Congress (2015-2016).

Termini, R. B., & Tressler, L. (2008). American beauty: An analytical view of the past and current effectiveness of cosmetic safety regulations and future direction. *Food & Drug Law Journal*, *63*(1), 257–274.

US Food and Drug Administration. (2015a). *Consumers: Cosmetics safety Q&A: Personal care products*. Retrieved from http://www.fda.gov/Cosmetics/ResourcesForYou/Consumers/ucm136560.htm (12/09/2015).

US Food and Drug Administration. (2015b). *Warning letters address drug claims made for products marketed as cosmetics. Guidance, compliance & regulatory information: Compliance and enforcement*. Retrieved from https://www.fda.gov/Cosmetics/ComplianceEnforcement/WarningLetters/ucm081086.htm.

US Food and Drug Administration. (2016a). *For industry: Color additives and cosmetics*. Retrieved from http://www.fda.gov/ForIndustry/ColorAdditives/ColorAdditivesinSpecificProducts/InCosmetics/ucm110032.htm.

US Food and Drug Administration. (2016b). *Laws & regulations—FDA authority over cosmetics: How cosmetics are not FDA-approved, but are FDA-regulated*. Retrieved from http://www.fda.gov/Cosmetics/GuidanceRegulation/LawsRegulations/ucm074162.htm.

US Food and Drug Administration. (2016c). FDA recall policy for cosmetics. Retrieved from http://www.fda.gov/Cosmetics/ComplianceEnforcement/RecallsAlerts/ucm173559.htm.

US Food and Drug Administration. (2016d). *Laws & regulations: Is it a cosmetic, a drug, or both? (or is it soap?)*. Retrieved from http://www.fda.gov/cosmetics/guidanceregulation/lawsregulations/ucm074201.htm.

U.S.C. (2010a). 21 U.S.C. § 321(g)(1).

U.S.C. (2010b). 21 U.S.C. § 321(i).

U.S.C. (2010c). 21 U.S.C. § 360(b).

U.S.C. (2016a). 15 U.S.C. § 1451.

U.S.C. (2016b). 21 U.S.C. § 361.

U.S.C. (2016c). 21 U.S.C. § 362.

CHAPTER

Tobacco products

11

Lilit Aladadyan*, Jonathan M. Samet**

**Tobacco Center of Regulatory Science (TCORS), University of Southern California, Los Angeles, CA, United States; **Colorado School of Public Health, Aurora, CO, United States*

"Cigars, cigarettes, and hookah tobacco are all smoked tobacco - addictive and deadly. We need effective action to protect our kids from struggling with a lifelong addiction to nicotine."

– Tom Frieden

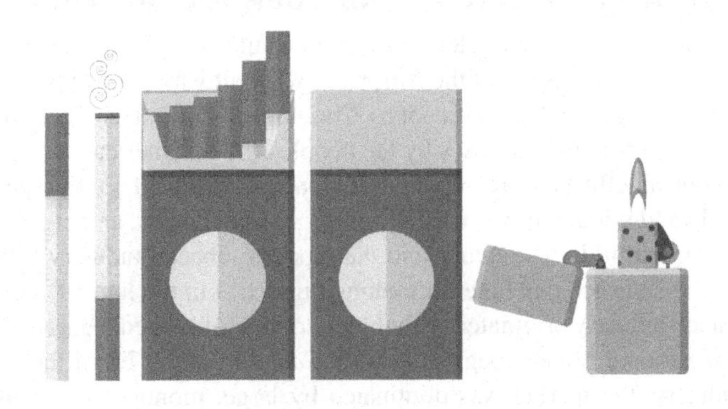

An Overview of FDA Regulated Products. http://dx.doi.org/10.1016/B978-0-12-811155-0.00011-9

CHAPTER OBJECTIVES

After reading this chapter, the reader will be able to:

- describe the history of tobacco and tobacco product regulation in the United States,
- introduce the Family Smoking Prevention and Tobacco Control Act (FSPTCA, TCA) and describe its major sections (such as Tobacco Product Standards, Modified Risk Tobacco Products),
- define what constitutes a tobacco product under the TCA and the Deeming Rule,
- describe the FDA Center for Tobacco Products (CTP) and its regulatory authorities,
- define the new "Public Health Standard" and illustrate how it differs from the traditional "Safety and Efficacy Standard," and
- describe the pathways through which tobacco products in the United States can be brought on to market.

11.1 HISTORY OF TOBACCO AND TOBACCO CONTROL

The use of tobacco products has a lengthy and colorful history that dates to the use of tobacco by the original peoples of the Americas, where it was a native plant unknown in Europe until brought to that continent by Christopher Columbus. Tobacco was primarily used for ceremonial purposes by the peoples of the Americas, and its potency and psychological effects were recognized. Once introduced to Europe, tobacco spread quickly and widely. It was used primarily in snuff, pipes, and cigars; cigarette use emerged in the mid-19th century and the modern tobacco industry followed the invention of the Bonsack machine for making cigarettes in the late 19th century. The modern tobacco industry originated with the rise of manufactured cigarettes and mass marketing of national brands, beginning with *Camel* in 1913. From the start of the cigarette industry, the market was dominated by large, monopolistic companies—American Tobacco founded by James B. Duke and R.J. Reynolds founded by Richard Joshua Reynolds. Cigarette sales rose quickly across the 20th century, first in men and later in women (Fig. 11.1) (U.S. Department of Health and Human Services, 2014).

Although there were earlier efforts to curb tobacco use, the modern era of tobacco control began around 1950 following findings from epidemiological studies that smoking was strongly associated with risk for lung cancer (U.S. Department of Health and Human Services, 2014). The first studies documenting this association were published by German scientists during the Nazi era of the 1940s. Beginning in the 1950s, there were numerous reports, initially from case–control studies and subsequently from prospective cohort studies, showing that smoking increased risk of lung cancer approximately 10-fold in smokers compared with never smokers. As the evidence mounted, the broad scientific community became increasingly convinced that cigarette smoking caused lung cancer and possibly other diseases, a conclusion that the tobacco industry attempted to counter through direct and indirect means, largely through creating doubt about the scientific evidence. Now, there is an extraordinarily large body of evidence causally linking active and passive smoking

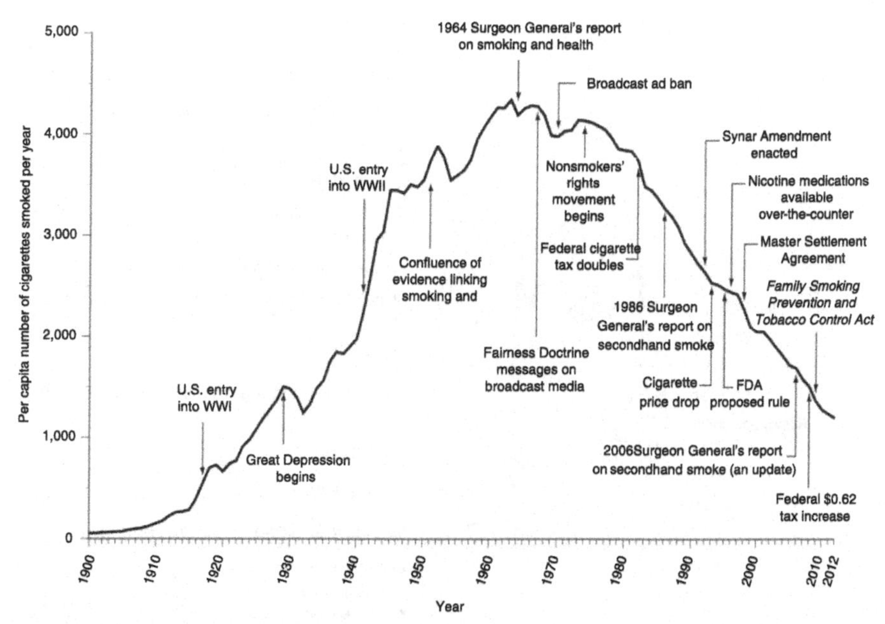

FIGURE 11.1 Adult per capita consumption and major smoking and health events, United States, 1900–2012.

to diseases of almost every organ of the body (Figs. 11.2 and 11.3) (U.S. Department of Health and Human Services, 2014).

With controversy mounting and mortality from lung cancer and other noncommunicable diseases rising, President Kennedy asked the Surgeon General to convene a committee to review the scientific evidence on smoking and health. The resulting 1964 report of the Advisory Committee to the Surgeon General stands as a landmark public health document. The report comprehensively reviewed the available evidence—thousands of papers, established a framework for causal inference, and reached the momentous conclusion that cigarette smoking caused lung cancer (in men) (U.S. Department of Health Education and Welfare, 1964). This and other findings of the report were sufficient to motivate action at national, state, and local levels. Fig. 11.1 captures some of the main tobacco control measures and documents the epidemic rise of cigarette smoking up to the 1960s and the subsequent sustained decline (U.S. Department of Health and Human Services, 2014).

11.2 HISTORY OF TOBACCO PRODUCT REGULATION

Regulatory activities have been central to tobacco control in the United States. At the federal level, the regulations have included the Surgeon General's warning on tobacco packs, restrictions of advertising, elimination of smoking on airplanes and other transportation environments, and banning of smoking in federal offices and

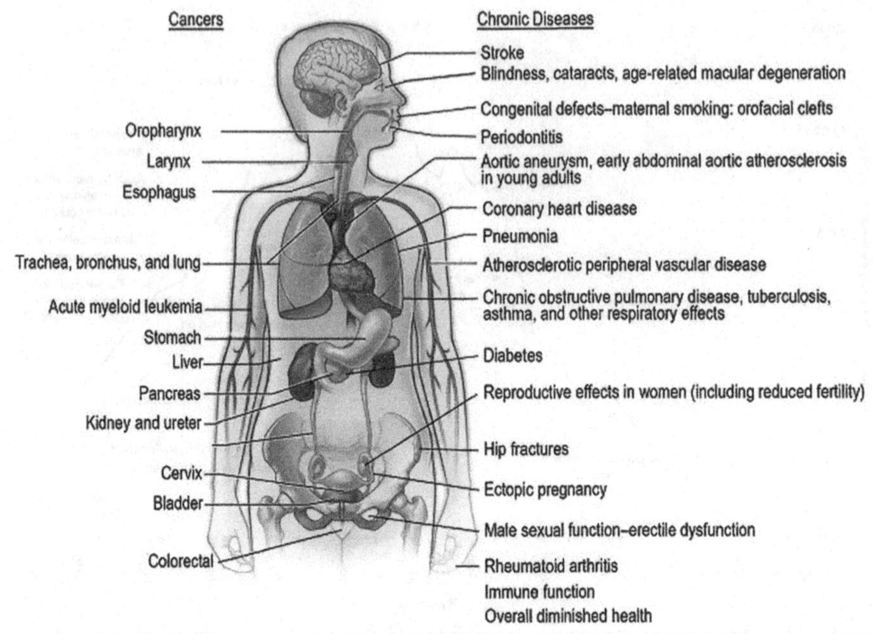

FIGURE 11.2 The health consequences causally linked to active smoking.

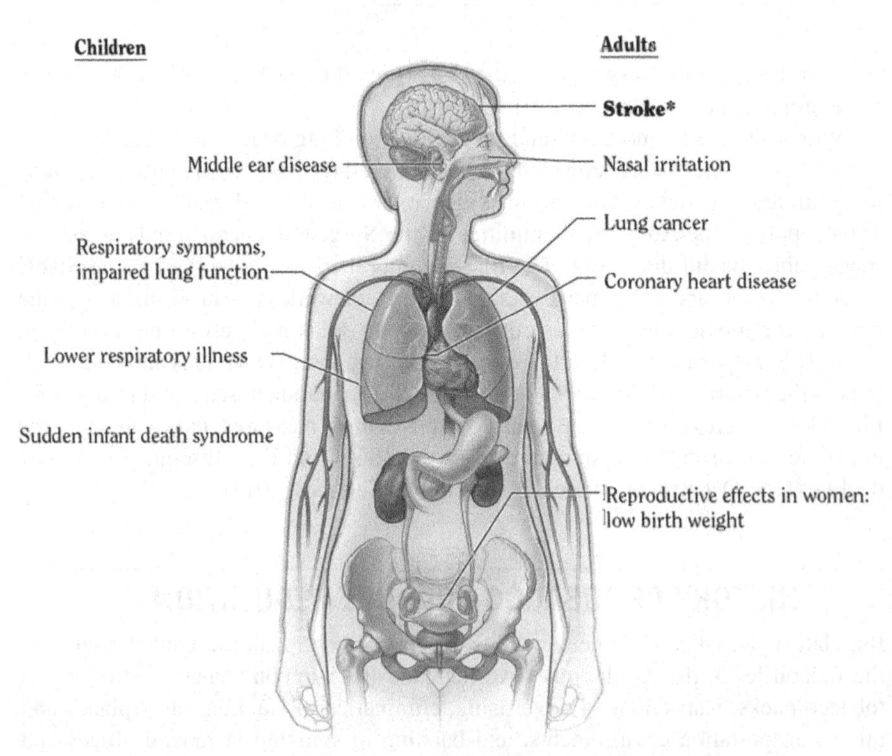

FIGURE 11.3 The health consequences causally linked to exposure to second-hand smoke.

most recently housing of the Housing & Urban Development Department. State and local ordinances have been the foundation for banning smoking in workplaces, bars, and restaurants and initially for restricting sales to minors.

The concept of broad federal regulation of tobacco products by the US Food and Drug Administration (FDA) originated in the 1990s, while Dr. David Kessler was the Commissioner. By then, tobacco products were causally linked to nicotine addiction and the experimentation that leads to addiction was known to begin in adolescence. Additionally, the tobacco industry had continued to deny the adverse health effects of smoking, was fraudulently misleading the public on the potential benefits of smoking lower-yield products ("light" or "mild"), and was aggressively attempting to counter the move to clean indoor air. Release of the industry's own documents confirmed its hidden knowledge of the adverse effects of its products, its misleading marketing, and its campaign to create doubt. Commissioner Kessler and his team viewed the moment as appropriate to move forward with regulation of the tobacco industry by the FDA.

In 1996, the FDA proposed its first tobacco regulation. But the United States Supreme Court concluded in 2000 that FDA lacks the authority to regulate tobacco products under the Food, Drugs, and Cosmetics Act (FDCA) (Food and Drug Administration, 2000). The 2009 passage of the FSPTCA (TCA) gave that authority to the FDA. While the origins of the TCA were controversial, it gave the FDA broad authorities over tobacco products and a new center, the CTP, to implement the new law. Since 2009, the FDA has undertaken a number of activities under its regulatory authority (Table 11.1). The TCA is intended to advance public health, and actions under the TCA are guided by their consequences on the public health. Generally, the FDA considers the consequences of its actions both for those using and those not using tobacco products. This population health impact standard or "public health standard" contrasts with the "safety and efficacy standard," typically applied to therapeutics and devices.

11.3 TOBACCO CONTROL ACT

On June 22, 2009, the FSPTCA (TCA) was signed into law, granting the FDA authority to regulate the manufacture, marketing, distribution, labeling, and sale of "tobacco products." The TCA amended the FDCA by the addition of Chapter IX, vesting the FDA with regulatory authority over tobacco products and creating the CTP to oversee the implementation of the TCA.

The TCA gave FDA immediate authority to regulate cigarettes, roll-your-own tobacco, and smokeless tobacco. In addition, it gave FDA the authority to use its rule-making process to extend its jurisdiction to other products it "deems" to be tobacco products like e-cigarettes, pipe tobacco, cigars, and hookah. The FDA proposed the rule, commonly known as the "Deeming Rule" in April 2014. The rule was finalized in May 2016 following a 105-day comment period and more than 130,000 comments.

Some of FDA's authorities under the law are to assess and collect user fees, impose advertising and promotion restrictions, impose warning labels, set performance standards and conduct postmarket surveillance, and review and approve (or disapprove) applications for new and modified risk tobacco products (MRTP; Table 11.1). Additionally, the FDA can require changes in the design and characteristics of both current and future tobacco products. These, and other, FDA CTP authorities are described in more detail in later sections of this chapter.

11.3.1 HOW ARE TOBACCO PRODUCTS DEFINED?

The definitions of tobacco products in the FSPTCA have proved critical as the FDA has moved forward with product regulations. The TCA defines a "tobacco product" as "any product made or derived from tobacco that is intended for human consumption, including any component, part, or accessory of a tobacco product (except for raw materials other than tobacco used in manufacturing a component, part, or

Table 11.1 Sections of the FSPTCA

Chapter IX – Tobacco Products	
SEC. 900	Definitions
SEC. 901	FDA authority over tobacco products
SEC. 902	Adulterated tobacco products
SEC. 903	Misbranded tobacco products
SEC. 904	Submission of health information to the Secretary
SEC. 905	Annual registration
SEC. 906	General provisions respecting control of tobacco products
SEC. 907	Tobacco product standards
SEC. 908	Notification and other remedies
SEC. 909	Records and reports on tobacco products
SEC. 910	Application for review of certain tobacco products
SEC. 911	Modified risk tobacco products
SEC. 912	Judicial review
SEC. 913	Equal treatment of retail outlets
SEC. 914	Jurisdiction of and coordination with the FTC
SEC. 915	Regulation requirement
SEC. 916	Preservation of state and local authority
SEC. 917	Tobacco Products Scientific Advisory Committee (TPSAC)
SEC. 918	Drug products used to treat tobacco dependence
SEC. 919	User fees

accessory of a tobacco product)" (Family Smoking Prevention and Tobacco Control Act, 2011). Components, parts, or accessories of tobacco products are included in the definition of "tobacco product" even when intended for further manufacturing into a finished consumer product. As an example, cigarette paper is regulated under the TCA as a "component" of a tobacco product; raw materials used to manufacture cigarette paper are not included in this definition.

11.3.2 DEEMING RULE

FDA published the final Deeming Rule in May 2016, extending its jurisdiction over all categories of products meeting the definition of "tobacco product" set forth in the FDCA, including Electronic Nicotine Delivery Systems (ENDS), cigars, hookah, pipe tobacco, nicotine gels, and any future tobacco products (Table 11.2). The rule applies to components and parts of deemed tobacco products like cartridges and atomizers but not to accessories like ashtrays and cigar clips. The Rule requires deemed products to comply with all provisions of the TCA and subjects subsets of these products, referred to as "covered" and/or "finished" tobacco products to additional restrictions. A "covered" tobacco product is a tobacco product, or component or part of a tobacco product that is itself made or derived from tobacco. For example, a bottle of e-liquid containing nicotine or electronic cigarettes prefilled with such an e-liquid are both covered products, whereas e-cigarette hardware sold separately from the e-liquid is not considered a covered product. A "finished" tobacco product, whether "covered" or not, is product that is in final packaging and intended for direct sale to consumers. Understanding these definitions helps to determine how different tobacco products are regulated. For example, only finished tobacco products are subject to premarket authorization.

Table 11.2 Regulated Tobacco Products

Products Under Immediate Regulatory Authority	"Deemed" Products
Cigarettes	Electronic Nicotine Delivery Systems (ENDS)
Cigarette tobacco	Pipe tobacco
Roll-your-own tobacco	Cigars
Smokeless tobacco	Hookah
Components, parts, and accessories of these products	E-liquid
	Any future products developed that meet the definition of "tobacco product" per the TCA
	Components and parts of these products
	Accessories of these products are not included

11.3.3 WHICH ENTITIES ARE SUBJECT TO FDA REGULATORY OVERSIGHT?

11.3.3.1 Tobacco product manufacturers

FDA has issued guidance to help clarify what is intended by the term "manufacturing" or "manufacturer." According to this guidance, any entity that manufactures, fabricates, assembles, processes, labels, or imports a tobacco product for sale or distribution in the United States is considered a manufacturer. The "deeming" rule further expanded this definition to include manufacturers of ENDS and e-liquids. Thus, any entity/establishment that mixes or prepares e-liquids or creates or modifies aerosolizing apparatus for direct sale to consumers is considered a manufacturer. This definition includes vape shops and other retail establishments.

11.3.3.2 Tobacco product distributors

A distributor is any entity who "furthers the distribution of a tobacco product, whether domestic or imported, at any point from the original place of manufacture to the person who sells or distributes the product to individuals for human consumption" (Section 900(7)).

11.3.3.3 Tobacco product retailers

A retailer is defined as any person or entity who sells tobacco products to individuals for personal consumption. This definition includes those entities that operate facilities with self-serve displays.

11.3.3.4 Tobacco growers, producers, and warehouses

FDA does not have jurisdiction to regulate activities/facilities of entities defined as growers, producers, or warehouses.

11.3.4 USER FEES

FDA CTP activities are funded with user fees collected from manufacturers and importers of tobacco products. Section 919 of the FDCA authorizes the FDA to collect these fees (predetermined by fiscal year), based on the class of tobacco product and its market share. The total amount of user fees that the CTP is authorized to assess reaches the proscribed maximum of $712 million in 2019. The fees are to be collected quarterly, with the quarterly amount to be allocated among the classes of tobacco products (including newly "deemed" products) based on the respective product class's volume of products in commerce. As per the TCA, these user fees may only be spent on tobacco product regulatory activities. Additionally, no other funds may be spent on FDA tobacco regulatory activities. Some of the user fees have been transferred to the National Institutes of Health (NIH) to support the Tobacco Regulatory Science Program (TRSP), which funds the Tobacco Centers of Regulatory Science (TCORS) and other research initiatives intended to support FDA activities.

11.4 CENTER FOR TOBACCO PRODUCTS

The TCA called for the establishment of the FDA CTP, to be solely financed through user fees collected from tobacco product manufacturers and importers. The CTP, which has now grown to over 600 staff across various offices (Fig. 11.4), is tasked with overseeing the implementation of the TCA. It has a highly multidisciplinary staff, reflecting the complexities of implementing science-based regulations and carrying out its broad public health mission.

11.4.1 CTP GOALS

Tobacco use is the leading cause of preventable death and disease in the United States (and across the world), causing an estimated 480,000 premature deaths yearly (U.S. Department of Health and Human Services, 2014). The goal of the FDA's CTP is to reduce this harm by (1) preventing people, especially youth, from starting to use tobacco products, (2) encouraging users to quit, and (3) reducing the adverse health impact for those who continue to use tobacco products Box 11.1.

11.4.2 KEY CTP ACTIVITIES

11.4.2.1 Compliance and enforcement

The CTP is responsible for monitoring industry compliance with the provisions of the TCA, including monitoring of manufacturers, distributors, importers, and retailers. The CTP has authority to issue warnings, civil penalties, seizures and injunctions, no-tobacco-sale orders, as well as the power to pursue criminal prosecution. This Office also helps industry to understand and comply with the law by issuing guidance documents and conducting training through webinars and conferences. How a product is classified is important in determining which center is responsible for its oversight. As can be seen in Box 11.2, a nicotine replacement therapy would be considered a drug and, hence, regulated by the Center for Drug Evaluation and Research (CDER).

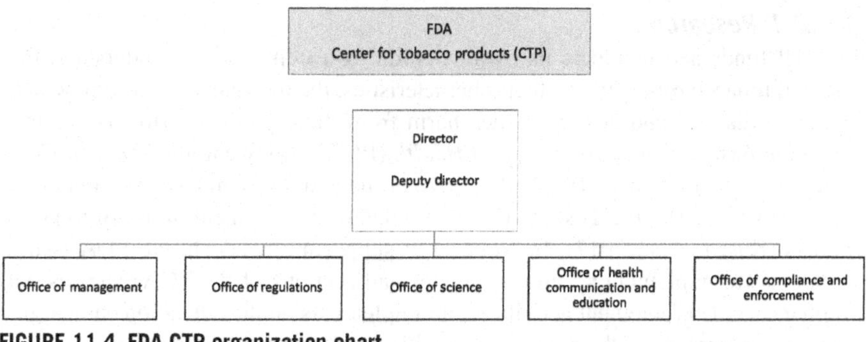

FIGURE 11.4 FDA CTP organization chart.

> **BOX 11.1 CTP MISSION**
>
> The mission of the CTP is "to protect Americans from tobacco-related death and disease by regulating the manufacture, distribution, and marketing of tobacco products and by educating the public, especially young people, about tobacco products and the dangers their use poses to themselves and others" (FDA CTP web site).

> **BOX 11.2 CTP VERSUS CDER**
>
> Products defined as a "drug," "device," or "combination product" under the FDCA are excluded from regulation by the CTP as a "tobacco product" and are instead regulated by the other respective FDA centers, such as the Center for Drug Evaluation and Research (CDER) or the Center for Devices and Radiological Health (CDRH). For example, nicotine replacement therapies such as nicotine gels, gums, or patches that are marketed as therapeutic agents to treat nicotine dependence are considered drugs and regulated by CDER; hence, a randomized clinical trial is required to establish efficacy.

11.4.2.2 Policy, rulemaking, and guidance

The CTP is responsible for setting policy and developing regulations and guidance to help operationalize the TCA and reduce the impact of tobacco use on the public health. The CTP Office of Regulations oversees the development of regulations from inception/proposed rule, to publication in the Federal Register, through the public comment period and publication of the final rule.

11.4.2.3 Public education campaigns

The CTP directly invests in public education campaigns to help educate the public, especially youth, about the dangers of tobacco products. These include advertising and print materials distributed through radio and TV, as well as digital media and displays at theme parks and malls. For example, "The Real Cost" campaign, released in 2014 to discourage youth tobacco use, was the first ever national public education campaign released by the FDA.

11.4.2.4 Research

The CTP funds and conducts its own scientific research to better understand the variety of tobacco products and their characteristics, the mechanisms through which they cause disease, and how to reduce harm from these products (Box 11.3). The *Population Assessment of Tobacco and Health* (PATH) study and the *Tobacco Centers of Regulatory Science* (TCORS) are two of the major research efforts supported by the FDA CTP. The PATH study provides a detailed assessment of use of tobacco products (Kasza et al., 2017). In addition to supporting research, the FDA issues regulations (to implement specific mandates and sections of the TCA), carries out compliance and enforcement activities, and implements public education campaigns to raise awareness about the dangers of smoking and tobacco product use.

> **BOX 11.3 CTP RESEARCH PRIORITIES**
>
> Product diversity—understanding the types of tobacco products and how their specific characteristics affect people's attitudes, beliefs, perceptions, and use of these products;
>
> Addiction—understanding what effect different levels of nicotine and other factors have on addiction;
>
> Toxicity and carcinogenicity—understanding how changes in tobacco products affect their potential for harm and ways to reduce that harm;
>
> Health consequences—understanding the risks of different tobacco products;
>
> Communication—finding ways to effectively convey information about the risks of using tobacco and about CTP's role in regulating tobacco products;
>
> Marketing—understanding the impact of tobacco product marketing and public education on people's attitudes, beliefs, perceptions, and use;
>
> Economics and policy—estimating the economic impact of CTP's regulations; also understanding how CTP's actions change tobacco use and illness and death from tobacco use.

11.5 TOBACCO PRODUCTS SCIENTIFIC ADVISORY COMMITTEE (TPSAC)

The TCA calls for the establishment of a 12-member advisory committee composed of members who are appointed based on technical qualification in diverse professional backgrounds. The 12-member committee is composed of nine voting members—seven individuals from the scientific community, one individual representing a local, state, or federal government, and one individual from the general public. The remaining three members are nonvoting members and represent the interests of the tobacco manufacturing industry, the small business tobacco manufacturing industry, and a representative of the tobacco growers. The make-up of TPSAC is unique in having participation from representatives of the tobacco industry. The role of TPSAC is to provide advice, information, and recommendations to the Secretary.

11.6 WHAT IS A "PUBLIC HEALTH STANDARD"?

FDA's traditional "safe and effective" standard for evaluating medical products does not apply to tobacco products. FDA evaluates new tobacco products based on a "public health standard" that considers the risks and benefits of the tobacco product to the population as a whole, including users and nonusers. Similarly, when developing certain regulations, the law requires FDA to apply a public health approach that considers the net effect (calculation of risks and benefits) of the regulatory action to the population as a whole—not just on individual users—with respect to initiation and cessation of tobacco use: the likelihood that existing users will stop using such tobacco products versus the likelihood that nonusers would start.

11.6.1 TOBACCO PRODUCT STANDARDS AND SPECIAL RULES

The FDA is authorized to establish tobacco product standards "appropriate for the protection of the public health." Such standards can be related to the composition, design, labeling, or marketing of tobacco products. Examples of potential standards, as identified in the TCA, may be related to product nicotine yields, reduction or elimination of harmful constituents or components, provisions for testing of tobacco products or for the measurement of product characteristics, among others. Where appropriate and allowable to the extent of the law, standards may also be adopted to restrict the sale and distribution of products, as well as require the use or prescribe the form and content of product labels. To determine the "appropriateness" of a standard, FDA must consider scientific evidence related to overall impact concerning (1) the risks and benefits to the population as a whole, including users and nonusers, (2) the likelihood that current users will stop using the product(s), and (3) the increased or decreased likelihood that those who do not use tobacco products will start using the product(s). Additionally, the FDA must consider the feasibility of industry compliance with the proposed standard, as well as other considerations, such as the potential creation of a demand for contraband products, etc.

Once adopted, product standards will undergo periodic reevaluations to determine whether the standard should be changed or revoked based on new scientific data. The adoption, amendment, and revocation of any standards must undergo the formal rule-making process (proposed rule, comment period, final rule). As can be seen in Box 11.4, FDA has limited authority regarding product standards.

As of the date of writing this chapter, the only example of an FDA proposed product standard is that for *N*-nitrosonornicotine (NNN) in finished smokeless tobacco products. The standard was proposed in January 2017 (Docket No. FDA-2016-N-2527) because NNN is a major contributor to the elevated oral cancer risk in users of smokeless tobacco products (U.S. Food and Drug Administration/Center for Tobacco Products, 2017). Thus, FDA found that limiting the mean level of NNN in finished smokeless tobacco would be appropriate for the protection of the public health. The proposed rule (proposed January 23, 2017) is open for public comments until July 10, 2017.

BOX 11.4 LIMITS TO FDA AUTHORITY WITH RESPECT TO PRODUCT STANDARDS

The Act expressly prohibits the FDA from establishing any standards that ban an entire category of products—such as cigarettes or cigars—and from setting standards that require the reduction of nicotine yields to zero. In July 2017, the FDA announced a new plan to shift its tobacco regulatory focus to nicotine and addiction to nicotine, with the goal of establishing product standards to lower nicotine levels in cigarettes to nonaddictive levels. The agency plans to issue an Advance Notice of Proposed Rulemaking (ANPRM) to seek public comments on the issue.

In addition to the authority to adopt new product standards, the TCA bans any artificial or natural flavor or any herb or spice other than tobacco and menthol that is a characterizing flavor (such as vanilla, strawberry, cherry, coffee, chocolate, etc.).

11.6.2 MENTHOL AS A FLAVOR-CHARACTERIZING ADDITIVE

Menthol is a natural compound found in certain plants as well as derived synthetically; it is widely used in many consumer products, including drugs and foods. When used directly in foods, it is a generally recognized as safe (GRAS) ingredient; when used in drugs or medicinal products, it is regulated as a drug with restrictions on its use and dosage. However, currently there are no product standards for its use in cigarettes. First added to cigarettes in the 1920s, mentholated cigarettes became widespread in the 1950s and 1960s and today account for more than a quarter of the US cigarette market (Federal Trade Commission, 2016). Due to its analgesic and cooling properties, when used as an additive in cigarettes, menthol reduces the harshness of the tobacco smoke and masks its bitter taste and irritating effects, making mentholated cigarettes more appealing, more addictive, and harder to quit (Tobacco Products Scientific Advisory Committee (TPSAC), 2011). The tobacco industry has been using flavorings to attract and addict youth and young adults to its products for decades, and mentholated cigarettes are inarguably the most pervasive and detrimental of all the flavored products. Congress's decision to exempt it from the cigarette flavorings ban was a political one and likely a concession to get industry support and help move the bill through the Senate. The decision was also controversial given that the disparities present in tobacco use are even higher with mentholated cigarette use. Instead, Congress delegated the controversial decision to the FDA and directed the TPSAC to prepare a report on the impact of mentholated cigarettes on the public health. The TPSAC delivered its report in March 2011, in which it concluded that the availability of menthol cigarettes has an adverse impact, and no benefit, on public health and recommended that removing menthol cigarettes from the marketplace would benefit public health (Tobacco Products Scientific Advisory Committee (TPSAC), 2011). The TPSAC report was followed by FDA's own independent evaluation of the existing research on menthol, which reached similar conclusions. However, to date, the FDA has not moved to implement a product standard for mentholated tobacco products, leaving it up to state and local governments to introduce policies addressing menthol.

11.6.3 MODIFIED RISK TOBACCO PRODUCTS

Prior to the passage of the TCA, tobacco companies were able to market products with a label indicating "light", "low", or "mild," leading consumers to believe that these products were safe or safer than other products on the market. In fact, these products did not lead to a reduction in health risks and may have actually increased tobacco use by increasing rates of initiation and reducing rates of quit attempts through false perceptions about safety or health risks. Under the TCA, products can have labels as MRTP if unique requirements are met along with marketing orders and postmarket monitoring conditions. Hence, smokeless tobacco products would not be considered as MRTPs (Box 11.5). These unique standards are to ensure that marketing of tobacco products does not mislead the public about health risks and that product claims are substantiated with scientific evidence.

Before companies can start marketing products with modified risk claims, they must file a Modified Risk Tobacco Product Application (MRTPA) and obtain one of two MRTP orders from the FDA—a *risk modification order* or an *exposure modification order* (Section 911(g))—in addition to satisfying any other applicable premarket review orders, such as those required of new tobacco products. FDA has issued a draft Guidance for Industry to assist with filing of MRTPAs (U.S. Food and Drug Administration/Center for Tobacco Products, 2012a). When reviewing applications for MRTP designations, FDA must consider the overall impact on public health, including the likelihood that smokers who would have otherwise quit will switch to MRTPs or to dual use of MRTPs with other products, or the likelihood that individuals who would not have otherwise initiated tobacco use will start using MRTPs due to them carrying less risk. The FDA will also consider the risks and benefits of the use of the MRTP compared to the use of existing FDA-approved smoking cessation products, which are regulated by CDER and CDRH (Box 11.6).

Except for trade secrets or other confidential proprietary information, the contents of an MRTPA will be made available to the public and FDA must consider any comments submitted to the docket before it can make a decision on the application. Additionally, the TCA requires that the FDA refer any MRTPAs to the Tobacco Products Scientific Advisory Committee (TPSAC) for recommendations prior to issuing a final decision on the application or issuing a marketing order. Once approved, MRTP orders are generally valid for a period of 5 years; however, manufacturers are required to carry out postmarket surveillance to determine the impact of the order on

BOX 11.5 SMOKELESS TOBACCO PRODUCTS

Smokeless tobacco products are a class of tobacco products and are not to be considered MRTPs simply because their label or advertising uses words or phrases such as "smokeless," "not consumed by smoking," "smokefree," "smoke-free," "does not product smoke," "no smoke," etc.

BOX 11.6 TOBACCO CESSATION PRODUCTS

Products that are intended to be used for treatment of tobacco dependence and cessation are not considered MRTPs. Tobacco cessation products, or products with such claims, will be considered as drug or device products and will be subject to the respective regulations established by the FDA Center for Drug Evaluation and Research (CDER) or the Center for Devices and Radiological Health (CDRH).

consumer perception, behavior, and health and to submit an annual report on the findings to the FDA. The FDA also maintains authority to withdraw a formerly issued authorization based on new findings. The first MRTPA was declined due to lack of evidence (Box 11.7).

11.7 HOW CAN TOBACCO PRODUCTS BE MARKETED?

11.7.1 TOBACCO PRODUCTS UNDER IMMEDIATE REGULATORY AUTHORITY AS PER THE TCA

A *new tobacco product* is defined by the TCA as any tobacco product (including those in test markets) that was not commercially marketed in the United States prior to February 16, 2007, or a product already on the market that was modified after that date. A tobacco product can be brought into market via one of the three pathways (Table 11.3).

If the product was in the market prior to February 16, 2007, it is defined as an "existing" or "grandfathered" product and can continue to be marketed in the United States if it remains unchanged. If a tobacco product was introduced between February 16, 2007 and March 21, 2011, the product is considered a "provisional new tobacco product" and requires the submission of a 905(j) (Substantial Equivalence) report. If the *substantial equivalence report* was submitted by the manufacturer for the new product prior to March 22, 2001, then the product can continue to be marketed unless found not substantially equivalent by the FDA. If a product is introduced to the market on or after March 22, 2011, or has been changed or altered, it would be defined as a "regular new tobacco product" and a marketing order from the FDA is required before it can be marketed. A marketing order can be obtained via three possible pathways: submission of a substantial equivalence report, submission of a premarket application, or submission of a substantial equivalence exemption report. In all three cases, an order from the FDA permitting marketing must be obtained before the product can be marketed in the United States.

Before the passing of the TCA, manufacturers could make changes to tobacco products without the knowledge of consumers or the public. Generally, product characteristics were altered in order to make the products more appealing, such as improving taste. Manufacturers are now prohibited from marketing new or modified tobacco products in the United States without receiving a marketing order from the FDA. The marketing order can be obtained by submitting an application for premarket review to the FDA (Section 910(a)(2)).

BOX 11.7 MRTPA FOR SNUS PRODUCTS

The first MRTPA submitted to the FDA was by Swedish Match North America, Inc. for 10 of its snus products. In December 2016, the FDA declined to issue a marketing order and encouraged the company to amend its application by altering the proposed labeling claim(s) and/or supplementing the evidence and conducting new studies.

Table 11.3 Requirements for Marketing Authorization for Cigarettes and Smokeless Tobacco

Product Category	Definition of Product Category	Submission(s) Required	Order Required
Grandfathered/ existing	Product was in the US market on or before February 15, 2007	None	None, can continue to market unless changes have been made to the product
Provisional new tobacco product	Product was introduced to the US market between February 16, 2007 and March 21, 2011	905(j) Report/ Substantial Equivalence Report	None if SER was submitted before March 22, 2001, unless product was found not substantially equivalent
New tobacco product	Product introduced to market after March 21, 2011	905(j) Report/ Substantial Equivalence Report Or PMTA (premarket tobacco application) Or SE exemption report	Yes. Marketing order required from FDA permitting marketing

11.7.2 NEWLY DEEMED TOBACCO PRODUCTS (CIGARS, E-CIGARETTES)

In August 2017, the FDA issued guidance on the extension of certain tobacco product compliance deadlines related to the final Deeming Rule. The (Table 11.4) below provides a summary of the new compliance dates for the various products affected.

11.8 SALE AND DISTRIBUTION

The FDA's "Sale and Distribution Rule" (21 CFR Part 1140), designed to protect children and adolescents, contains provisions that restrict the sale and distribution of tobacco products. For example, the Rule restricts manufacturers, distributors, and retailers from distributing free samples, sponsoring events, or providing promotional items and services. The Rule also contained restrictions on advertising, but these were found to be unconstitutional under the First Amendment, and hence the FDA does not currently enforce these (Commonwealth Brands, 2010). The Rule initially applied to cigarettes and smokeless tobacco products but was amended with the Deeming Rule to include all products meeting the statutory definition of tobacco product.

Table 11.4 Premarket Review Requirements

Submission Type	Products Affected	Revised Compliance Date
Substantial equivalence exemption request; substantial equivalence report; premarket tobacco product application (PMTA)	New tobacco products[a]; Newly deemed finished tobacco products on market as of August 8, 2016	August 8, 2021 (combustible products); August 8, 2022 (noncombustible products)

[a]*A "new tobacco product" is any tobacco product, including those in test markets, that was not commercially marketed in the United States as of February 15, 2007, or any modification of a tobacco product where the modified product was commercially marketed in the United States after February 15, 2007.*

11.9 OTHER REGULATORY REQUIREMENTS

11.9.1 DISCLOSURE OF INGREDIENTS

According to Section 904(a)(1) of the TCA, companies are required to provide a full list of ingredients for all regulated products. A guidance document was issued to assist industry with compliance (U.S. Food and Drug Administration/Center for Tobacco Products, 2009).

11.9.2 DISCLOSURE OF HARMFUL AND POTENTIALLY HARMFUL INGREDIENTS

As directed by the TCA, the FDA has established a current list of 93 harmful and potentially harmful ingredients (HPHCs) present in tobacco products. Manufacturers are required to measure and report to the FDA the levels of HPHCs present in their products by brand and subbrand. The FDA has published a guidance to help industry with compliance and reporting of HPHCs (U.S. Food and Drug Administration/Center for Tobacco Products, 2012b). The FDA is required to publicly disclose the information it gathers about HPHCs in a way that is understandable and not misleading.

11.9.3 REGISTRATION AND PRODUCT LISTING

As per Section 905 of the TCA, companies must register manufacturing facilities and provide a list all regulated products. A guidance document was issued to assist industry with compliance (U.S. Food and Drug Administration/Center for Tobacco Products, 2016).

11.9.4 ADULTERATION AND MISBRANDING (SECTIONS 902 AND 903)

Similar to foods, drugs, and devices, an adulterated or misbranded tobacco product cannot be introduced into interstate commerce. A tobacco product is considered "adulterated" if any of eight conditions apply. Four of these conditions address the same standards of quality applied to other FDA-regulated products and address the composition of the product itself, its packaging, the conditions under which the product has been prepared and stored, as well as adherence to established good manufacturing practice requirements. The other conditions are more specific to tobacco products and include violations of MRTP requirements or marketing authorizations, nonconformance to tobacco product standards, or failure to pay assessed user fees.

The TCA also lists conditions under which a tobacco product can be deemed "misbranded" and applies to product labeling, package, and advertising. For example, a product's labeling or advertising cannot be false or misleading and must adhere to the labeling and advertising requirements set forth in the TCA (Sections 903 and 906); the package must bear all applicable required information, such as the percentage of domestically and foreign-grown tobacco contained in the package; and the product must meet all registration, product listing, and reporting requirements.

11.10 CONCLUSIONS

With the passage of the Family Smoking Prevention and Tobacco Control Act in 2009, the United States took a major leap forward in its efforts to combat tobacco use and reduce the toll of death and disease attributable to tobacco smoke. For the first time, the United States enabled a regulatory agency, the FDA, to regulate the use of tobacco by consumers and the manufacture, sale, and advertising of it by industry. The charge was a formidable one—develop and implement regulations that are consistent with the TCA, are grounded in science, and consistent with limitations imposed by countless other laws, including the US Constitution. To do so, the FDA, never before having had regulatory authority over tobacco products, had to build a regulatory body and infrastructure from the ground up. Fast forward to 2017, and while frustrations abound about the lack of FDA action on some important matters (such as inaction on mentholated cigarettes), the agency has, nevertheless, had other important accomplishments (such as the passing of the Deeming Rule).

There are challenges that are unique to the regulation of tobacco products. As opposed to other FDA-regulated products, tobacco is inherently dangerous and the individual-level safety and efficacy standards that the FDA has historically applied to products under its regulatory authority cannot be applied to tobacco products. Instead, the FDA has had to develop a new public health standard, a population-level assessment of *likely* impact on overall public health. Additionally, the FDA has had to develop and validate entirely new procedures for testing and measurements of constituents in tobacco and tobacco smoke (such as for HPHCs), it has had to develop entirely new procedures and metrics for the evaluation of product submissions (i.e. substantial equivalence, premarket tobacco applications, etc.) and has had to face an unprecedented level of legal challenges brought on by industry (i.e. challenges to the Deeming Rule, warning labels on cigarette packs, etc.).

Despite the multitude of challenges, the FDA has managed to publish dozens of rules (i.e. the Deeming Rule, procedural rules, rules establishing levels of HPHCs, etc.) and guidance documents, review and make decisions on thousands of marketing applications, conduct compliance and inspection activities, and, among other things, directly engage in scientific research and public education campaigns. In the overall tobacco control landscape, local and state governments continue to outpace the FDA with novel laws aimed at reducing tobacco use and the associated disease burden, such as raising the minimum of legal access to tobacco products, restricting the sale of flavored products, cigarette tax increases, public smoking restrictions, restricting sales in pharmacies, etc. The FDA, however, now holds the responsibility for creating macrolevel change.

ADDENDUM/POSTSCRIPT

In July 2017, FDA Commissioner Dr. Scott Gottlieb unveiled the agency's new and comprehensive plan for regulation of tobacco products. The cornerstone of the new plan is regulation of nicotine via product standards that reduce the amount of nicotine in cigarettes to nonaddictive levels. The agency's new plan also focuses on flavors in tobacco products and the regulation of flavors that are particularly appealing to youth. Following the announcement, in March 2018, the agency issued an Advanced Notice of Proposed Rulemaking (ANPRM) to obtain public comments on the development of a potential nicotine product standard for combustible cigarettes. It issued a second ANPRM to obtain information related to how flavors attract youth to initiate tobacco product use and whether certain flavors can promote harm reduction in adult smokers by helping them switch from combustible to noncombustible tobacco products.

Additionally, the new plan extends the deadline for Premarket Review Applications for newly "deemed" products. Applications for combustible products, including cigars, pipe tobacco and hookah tobacco that were on the market as of August 8, 2016, will be due by August 8, 2021. Applications for noncombustible products, such as e-cigarettes and other vapor products, will be due by August 8, 2022. Manufacturers will be able to continue to market products while the agency reviews product applications.

REFERENCES

Commonwealth Brands. (2010). Commonwealth Brands, Inc. v. United States, 678 F. Supp. 2d 512. (W.D. Ky. 2010).

Family Smoking Prevention and Tobacco Control Act. (2011). Family Smoking Prevention and Tobacco Control Act, 21 U.S.C. § 387.

Federal Trade Commission. (2016). *Cigarette report for 2014.* <https://www.ftc.gov/system/files/documents/reports/federal-trade-commission-cigarette-report-2014-federal-trade-commission-smokeless-tobacco-report/ftc_cigarette_report_2014.pdf>

Food and Drug Administration. (2000). Food and Drug Administration, et al. v. Brown & Williamson Tobacco Corp., et al., 529 U.S. 120. (United States Supreme Court).

Kasza, K. A., Ambrose, B. K., Conway, K. P., Borek, N., Taylor, K., Goniewicz, M. L., et al. (2017). Tobacco-product use by adults and youths in the United States in 2013 and 2014. *New England Journal of Medicine*, *376*(4), 342–353. doi: 10.1056/NEJMsa1607538.

Tobacco Products Scientific Advisory Committee (TPSAC). (2011). *Menthol cigarettes and public health: Review of the scientific evidence and recommendations.* <https://www.fda.gov/downloads/advisorycommittees/committeesmeetingmaterials/tobaccoproductsscientificadvisorycommittee/ucm269697.pdf>

U.S. Department of Health and Human Services. (2014). *The health consequences of smoking: 50 years of progress. A report of the Surgeon General.* <http://www.ncbi.nlm.nih.gov/books/n/surgsmoke50/pdf/>

U.S. Department of Health Education and Welfare. (1964). *Smoking and health: Report of the Advisory Committee to the Surgeon General of the Public Health Service (Public Health Service Publication No. 1103).* <https://profiles.nlm.nih.gov/ps/access/NNBBMQ.pdf>

U.S. Food and Drug Administration/Center for Tobacco Products. (2017). *Tobacco product standard for N-nitrosonornicotine level in finished smokeless tobacco products*. (Docket No. FDA-2016-N-2527). <https://www.fda.gov/downloads/AboutFDA/ReportsManualsForms/Reports/EconomicAnalyses/UCM537872.pdf>

U.S. Food and Drug Administration/Center for Tobacco Products. (2012a). *Draft guidance for industry: Modified risk tobacco product applications.* <https://www.fda.gov/downloads/TobaccoProducts/Labeling/RulesRegulationsGuidance/UCM297751.pdf>

U.S. Food and Drug Administration/Center for Tobacco Products. (2012b). *Draft guidance for industry: Reporting harmful and potentially harmful constituents in tobacco products and tobacco smoke under Section 904(a)(3) of the Federal Food, Drug, and Cosmetic Act.*

U.S. Food and Drug Administration/Center for Tobacco Products. (2009). *Guidance for industry: Listing of ingredients in tobacco products.* <https://www.fda.gov/downloads/TobaccoProducts/GuidanceComplianceRegulatoryInformation/ucm192053.pdf>

U.S. Food and Drug Administration/Center for Tobacco Products. (2016). *Revised guidance for industry: Registration and product listing for owners and operators of domestic tobacco product establishments.*

CHAPTER

Quality

12

Michael Jamieson, Nancy Pire-Smerkanich
International Center for Regulatory Science, University of Southern California, Los Angeles, CA, United States

"Quality is never an accident. It is always the result of intelligent effort."
– John Ruskin

An Overview of FDA Regulated Products. http://dx.doi.org/10.1016/B978-0-12-811155-0.00012-0

CHAPTER OBJECTIVES

After reading this chapter, the reader will be able to:

- identify the three best practices that companies must follow to ensure the quality of the products that they manufacture,
- recognize where in the development cycle of a pharmaceutical these best practices apply,
- understand the basic principle of GLPs, GCPs, and GMPs, and
- describe how a drug–device combination product is regulated by the FDA.

12.1 BACKGROUND

One of the main responsibilities of regulatory agencies around the globe is to ensure the quality of all healthcare products that are marketed in their respective countries. To do this, industry and regulators have developed, and individual countries have endorsed, two overarching systems that work in parallel; "good practices", often referred to as GxPs, and quality systems. The three main GxPs are good laboratory practices (GLPs), good clinical practices (GCPs), and good manufacturing practices (GMPs). These three GxPs coincide with the various stages of development that most healthcare products go through. GLPs cover the nonclinical laboratory testing, such as animal studies, and bench testing that are submitted to the regulatory agencies in support of a clinical trial application and marketing authorization. GCPs cover the design, conduct, performance, monitoring, auditing, recording, analysis, and reporting of the clinical trials that will be used to support an application for marketing authorization. GMPs cover the manufacturing, testing, and quality assurance that ensure that a product is safe for human use. Even though some countries have their own version of some, or all of these three GxPs, for traditional medical devices, pharmaceuticals, and biologics, these three GxPs are very similar across the globe.

When the International Council for Harmonization (ICH) was developing the Technical Requirements for Pharmaceuticals for Human Use (actually the full and formal name of ICH), it coined three important terms for the scientific and technical guidelines it would focus on for the purpose of drug and biologics registration: Quality (Q), safety (S), and efficacy (E) (www.ich.org). These disciplines align with the sections of the dossiers associated with chemistry, manufacturing, and controls (CMC), preclinical, and clinical. The mission of the ICH regulatory authorities and industry members was, and is, to achieve greater harmonization worldwide to ensure that safe, effective, and high-quality medicines are developed and registered in the most resource-efficient manner.

It is important to note that these terms were never meant to be mutually exclusive and that quality, taken as a feature or attribute, needs to be associated with manufacturing process as well for both preclinical and clinical studies.

12.2 GOOD LABORATORY PRACTICES FOR NONCLINICAL LABORATORY STUDIES

GLP regulations, enacted in the United States in the late 1970s, are intended to ensure that investigators conduct safety and efficacy studies in a controlled, documented, and traceable manner. The term GLP is most commonly associated with the pharmaceutical industry and the required nonclinical animal testing that must be performed prior to approval of new drug products. However, GLP applies to many other nonpharmaceutical agents such as medical devices and biologics. These regulations are intended to assure the quality and integrity of the data necessary to support FDA investigational new d rug (IND) and investigational device exemption (IDE) applications for first-in-human investigations, as well as for subsequent FDA marketing approval. In the United States, the GLP regulations can be found in Code of Federal Regulations (21 CFR 58). The US regulations define "nonclinical laboratory studies as 'in vivo or in vitro' experiments in which test articles are studied prospectively in test systems under laboratory conditions to determine their safety." It is important to understand that GLP defines a set of quality standards for the conduct of nonclinical studies, data collection, and the reporting of the results. GLP does not do anything to ensure that the research being conducted would be considered "good science."

To fully understand the scope of what is covered under GLP, it is probably easiest to start with a look at all of the sections that are included under 21 CFR 58 (Table 12.1).

One of the most important concepts contained within the GLPs is the idea of a "quality unit (QA)." This QA unit is intended to be an independent group or individual that monitors the entire study conduct, analysis, and reporting. One of the main ways the QA unit does this is through the auditing of standard operating procedures (SOPs) to ensure that the researchers are actually following the approved protocol and procedures used in conducting the clinical trial.

The US GLP regulations are not the only GLP regulations that are used by regulators around the world. Many other countries, including the European Union, follow the Organization for Economic Co-operation and Development (OECD) principles of good laboratory practice.

12.3 GOOD CLINICAL PRACTICES

In the US FDA regulations, there is no single section or subsection designated for GCPs. Instead, the FDA has adopted and issued a Guidance for Industry on GCP, which is the ICH E6 good clinical practice: Consolidated guidance (1996). This guideline, particularly the second revision (R2), addresses multiple sections of the US regulations (Title 21 Food and Drugs): Part 11 (Electronic records and signatures); Parts 50 (Protection of Human Subject); Part 54 (Financial Disclosure by Clinical Investigators); Part 56 (Institutional Review Boards); Part 312 (Investigational new drug application); Part 314 (Applications for FDA Approval to Market a

Table 12.1 Good Laboratory Practices for Nonclinical Laboratory Studies (21 CFR 58)

Subparts 21 CFR 58	
Subpart A—General Provisions	§ 58.1—Scope § 58.3—Definitions § 58.10—Applicability to studies performed under grants and contracts § 58.15—Inspection of a testing facility
Subpart B—Organization and Personnel	§ 58.29—Personnel § 58.31—Testing facility management § 58.33—Study director § 58.35—Quality assurance unit
Subpart C—Facilities	§ 58.41—General § 58.43—Animal care facilities § 58.45—Animal supply facilities § 58.47—Facilities for handling test and control articles § 58.49—Laboratory operation areas § 58.51—Specimen and data storage facilities
Subpart D—Equipment	§ 58.61—Equipment design § 58.63—Maintenance and calibration of equipment
Subpart E—Testing Facilities Operation	§ 58.81—Standard operating procedures § 58.83—Reagents and solutions § 58.90—Animal care
Subpart F—Test and Control Articles	§ 58.105—Test and control article characterization § 58.107—Test and control article handling § 58.113—Mixtures of articles with carriers
Subpart G—Protocol for and Conduct of a Nonclinical Laboratory Study	§ 58.120—Protocol § 58.130—Conduct of a nonclinical laboratory study
Subpart H— [Reserved]	
Subpart J—Records and Reports	§ 58.185—Reporting of nonclinical laboratory study results § 58.190—Storage and retrieval of records and data § 58.195—Retention of records
Subpart K—Disqualification of Testing Facilities	§ 58.200—Purpose § 58.202—Grounds for disqualification § 58.204—Notice of and opportunity for hearing on proposed disqualification § 58.206—Final order on disqualification § 58.210—Actions upon disqualification § 58.213—Public disclosure of information regarding disqualification § 58.215—Alternative or additional actions to disqualification § 58.217—Suspension or termination of a testing facility by a sponsor § 58.219—Reinstatement of a disqualified testing facility

New Drug); Part 601 (Applications for FDA Approval of a Biologic License); Part 812 (Investigational Device Exemption); and Part 814 (Premarket Approval of Medical Devices) (https://www.fda.gov/downloads/Drugs/Guidances/UCM464506.pdf).

As stated clearly in the Introduction of the ICH E6 (R2) guideline, a GCP is an international ethical and scientific quality standard for designing, conducting, recording, and reporting trials that involve the participation of human subjects (www.ich.org). One of the goals of GCP is to assure the quality, reliability, and integrity of the data collected. It goes on to illustrate the sponsor's role of quality control and quality assurance and as elements of the clinical study protocol, as well as describing how essential documents are needed to permit evaluation of the conduct of a study and the quality of the data produced. This responsibility for the quality and integrity of the trial data ultimately resides with the sponsor, even when trials or functions are outsourced to Contract Research Organizations (CRO).

The question then becomes how sponsors ensure that there is high-quality data being collected in clinical studies. That responsibility is described to include quality control and quality assurance. Although these terms seem to indicate two different "job descriptions," both processes should be integrated into each study team member's duty. Sponsor companies need to establish systems with procedures that assure the quality of every aspect of the trial, beginning with the way the data is collected and monitored at the site and continuing throughout the study, until a final study report is generated.

In E6, ICH has defined these activities as follows:

- Quality control (QC): The operational techniques and activities undertaken within the quality assurance system to verify that the requirements for quality of the trial-related activities have been fulfilled.
- Quality assurance (QA): All those planned and systematic actions that are established to ensure that the trial is performed and the data are generated, documented (recorded), and reported in compliance with GCPs and the applicable regulatory requirement(s).

How these activities are operationalized may mean ensuring that there is an ethics committee available to review and approve the protocol and informed consent form. Additionally, oversight is needed to ensure that the data is entered correctly on case report forms at the site or sponsor and to ensure that the site is monitored by trained and qualified individuals with proper control from the sponsor. An independent evaluation of these activities may also be performed through QA audits of sites to determine whether standard operating procedures (SOPs) are being followed, computer systems are validated, and sponsor and site staff are adequately trained.

The recent revision to E6 places an increased emphasis on the use of quality systems and quality management systems in clinical trials. To differentiate the two, it is helpful to think of a quality system as one that focuses on quality practices, such as written SOPs, while a quality management system focuses on product and service quality and the means to achieve it. Quality management uses quality assurance practices and controls of these processes to achieve a high standard for clinical trial conduct.

For further information on quality management systems, one can refer to a standard published by the International Organization for Standardization (www.iso.org), an international standards writing body, which published ISO 9001, a highly regarded and well-known international standard for quality management.

It is important to remember that the voluntary nature of subject participation in clinical trials is one that should be honored by effectively and efficiently utilizing the data collected in studies. This process can only occur when high quality data is collected, analyzed, and summarized; and the integrity of this data is carefully maintained.

12.4 GOOD MANUFACTURING PRACTICES

In simple terms, GMPs provide guidelines for manufacturing, testing, and quality assurance to ensure that a product is safe for human use. In the United States, the main GMP that serves as a framework for the GMPs for different product types (such as compounding facilities or manufacturers of dietary supplements, for example) is titled "Current Good Manufacturing Practice for Finished Pharmaceuticals" (21 CFR 211). The requirements incorporated in this document, codified first in the Federal Register and then in the Code of Federal Regulations (21 CFR 211), are considered to belong to a "regulation," an expansion of the intentions of the Food Drug and Cosmetic Act, with which manufacturers must comply or are considered to be in violation of the law. A large number of countries also have their own GMP regulations, but when you put them all side-by-side for comparison, there really is not a significant number of differences between them (Table 12.2).

For someone new in the field, it is not always easy to understand the distinction between a quality system and GMP; in reality, the line is blurred even for people that have been in industry for years. Quality systems evolved as an expansion of GMPs to cover areas of activity included in GMPs, but QSRs go further, to encompass activities beyond manufacturing. International quality standards like the ones developed by the International Standards Organization (ISO) often cover most of what is included in GMPs, but in many cases these quality standards are more encompassing than GMPs and often have a lower level of detail. For example, the ICH Q10 Pharmaceutical Quality Systems guidance covers everything from the development of the drug right through to the disposal of the drug, whereas the US GMP only covers manufacturing of the product after it has been approved for commercialization. Some in industry would even argue that GMP, in itself, is a quality system, one that is specific to the manufacturing of the product. Knowing the difference between GMP and quality systems is really not that important, rather what is important is understanding that the two work in parallel to help ensure the quality of all healthcare products.

You will often see the acronym cGMP rather than just GMP and people will often ask what the difference between the two is. The "c" stands for "current" and refers to compliance with the most current GMP regulations.

Even though the international GMPs are very similar to each other, there are still many countries, including the United States, that require companies that want to sell healthcare products into their country to comply with that country's own GMP.

Table 12.2 Internationally Recognized GMPs and Quality Systems

Good Manufacturing Practices (GMPs) and Quality Systems	Source	Reference
Current Good Manufacturing Practices for Finished Pharmaceuticals	US FDA	21 CFR 211
General Biological Product Standards	US FDA	21 CFR 610
Quality System Regulations (Medical Devices)	US FDA	21 CFR 820
Good Manufacturing Practice	ICH	ICH Q7
Pharmaceutical Quality Systems	ICH	ICH Q10
Quality of Biotechnological Products	ICH	ICH Q5
Medical Devices; Quality Management Systems	ISO	ISO 13485
Good Manufacturing Practice	WHO	WHO

12.4.1 PHARMACEUTICALS

To fully understand the scope of what is covered under the pharmaceutical GMPs, it is probably the easiest to start with a look at all of the sections that are included (Table 12.3).

As we described earlier in this chapter, the quality of pharmaceuticals relies on a number of systems working together. According to the World Health Organization (WHO):

> The concepts of quality assurance, GMP, quality control and quality risk management are interrelated aspects of quality management, and should be the responsibility of all personnel.

12.4.2 MEDICAL DEVICES

As can be seen in Table 12.2, there are two major quality systems for medical devices in the world: (i) The US FDA's Quality System Requirements (QSRs) found in 21 CFR 820 and (ii) The International Standards Organization's (ISO) ISO 13485 2016.

The FDA defines medical device quality as "the totality of features and characteristics that bear on the ability of a device to satisfy fitness-for-use, including safety and performance." ISO defines medical device quality as "the totality of features and characteristics of a product or service that bear on its ability to satisfy stated or implied needs." Basically, these quality systems outline the methods used in, and the facilities and controls used for, the design, manufacture, packaging, labeling, storage, installation, and servicing of all finished devices intended for human use. The requirements contained in these quality systems are intended to ensure that finished devices will be safe and effective.

Before a device is commercialized, design engineers must transfer the design outputs to a production group, who then establish a "device master record" that specifies how the product will be manufactured under conditions that assure quality and consistency. Thus, a key activity from the point at which clinical trials are considered is that of freezing the design and putting into place the equipment and methods to

Table 12.3 Current Good Manufacturing Practices for Finished Pharmaceuticals (21 CFR 211)

Subparts 21 CFR 211	
Subpart A—General Provisions	§ 211.1—Scope
	§ 211.3—Definitions
Subpart B—Organization and Personnel	§ 211.22—Responsibilities of quality control unit
	§ 211.25—Personnel qualifications
	§ 211.28—Personnel responsibilities
	§ 211.34—Consultants
Subpart C—Buildings and Facilities	§ 211.42—Design and construction features
	§ 211.44—Lighting
	§ 211.46—Ventilation, air filtration, air heating and cooling
	§ 211.48—Plumbing
	§ 211.50—Sewage and refuse
	§ 211.52—Washing and toilet facilities
	§ 211.56—Sanitation
	§ 211.58—Maintenance
Subpart D—Equipment	§ 211.63—Equipment design, size, and location
	§ 211.65—Equipment construction
	§ 211.67—Equipment cleaning and maintenance
	§ 211.68—Automatic, mechanical, and electronic equipment
	§ 211.72—Filters
Subpart E—Control of Components and Drug Product Containers and Closures	§ 211.80—General requirements
	§ 211.82—Receipt and storage of untested components, drug product containers, and closures
	§ 211.84—Testing and approval or rejection of components, drug product containers, and closures
	§ 211.86—Use of approved components, drug product containers, and closures
	§ 211.87—Retesting of approved components, drug product containers, and closures
	§ 211.89—Rejected components, drug product containers, and closures
	§ 211.94—Drug product containers and closures
Subpart F—Production and Process Controls	§ 211.100—Written procedures; deviations
	§ 211.101—Charge-in of components
	§ 211.103—Calculation of yield
	§ 211.105—Equipment identification
	§ 211.110—Sampling and testing of in-process materials and drug products
	§ 211.111—Time limitations on production
	§ 211.113—Control of microbiological contamination
	§ 211.115—Reprocessing

Table 12.3 Current Good Manufacturing Practices for Finished Pharmaceuticals (21 CFR 211) (*cont.*)

Subparts 21 CFR 211	
Subpart G—Packaging and Labeling Control	§ 211.122—Materials examination and usage criteria
	§ 211.125—Labeling issuance
	§ 211.130—Packaging and labeling operations
	§ 211.132—Tamper-evident packaging requirements for over-the-counter (OTC) human drug products
	§ 211.134—Drug product inspection
	§ 211.137—Expiration dating
Subpart H—Holding and Distribution	§ 211.142—Warehousing procedures
	§ 211.150—Distribution procedures
Subpart I—Laboratory Controls	§ 211.160—General requirements
	§ 211.165—Testing and release for distribution
	§ 211.166—Stability testing
	§ 211.167—Special testing requirements
	§ 211.170—Reserve samples
	§ 211.173—Laboratory animals
	§ 211.176—Penicillin contamination
Subpart J—Records and Reports	§ 211.180—General requirements
	§ 211.182—Equipment cleaning and use log
	§ 211.184—Component, drug product container, closure, and labeling records
	§ 211.186—Master production and control records
	§ 211.188—Batch production and control records
	§ 211.192—Production record review
	§ 211.194—Laboratory records
	§ 211.196—Distribution records
	§ 211.198—Complaint files
Subpart K—Returned and Salvaged Drug Products	§ 211.204—Returned drug products
	§ 211.208—Drug product salvaging

manufacture the product in larger quantities. At the stage of early feasibility studies, sufficient attention to quality assurance must be paid so that patients are not put at risk. However, it may not be possible or financially justifiable to establish a production line until early clinical trials have convinced the company that the product design can be frozen. By the pivotal trial stage, the product to be tested should be in the form that it will be when it is commercialized. Key bench and animal tests should also be conducted on product that is in its final form and has been manufactured under quality systems regulations.

In addition to design controls, QSRs also include extensive and rigorous rules regarding the conduct and management of manufacturing. FDA will inspect the company to assure its compliance with these rules. Because different products are manufactured in different ways, each company must develop its own standard operating procedures and validation plans to govern manufacturing processes and equipment, as well as activities such as those related to component and reagent acquisition, product packaging and distribution that might affect product quality. At the beginning of pivotal trials, it is expected that the foundational elements of the quality system will be quite mature, but the equipment used for manufacture may not be the same as that which will be needed when processes are scaled up to increase the volumes of product needed after commercialization. Comparability arguments and even laboratory studies may have to be conducted to demonstrate that the product proposed for commercialization is equivalent to the product that was tested in clinical trials. Otherwise, the FDA may require that clinical trials be repeated on the new form of the device.

12.4.3 COMBINATION PRODUCTS

The FDA defines a Combination Product as "a product composed of any combination of a drug and a device; a biological product and a device; a drug and a biological product; or a drug, device, and a biological product." Most of us envision a combination product as something like an auto injector where the drug to be delivered is preloaded into an auto injector system and the product is marketed and sold as a single unit. This is indeed a combination product, but the FDA also recognizes other configurations of drug, devices, and biologics as combination products too.

The FDA has broken down combination products into four groups (21CFR Part 4):

1. A product comprising two or more regulated components, that is, drug/device, biologic/device, drug/biologic, or drug/device/biologic, that are physically, chemically, or otherwise combined or mixed and produced as a single entity;
2. Two or more separate products packaged together in a single package or as a unit and comprised of drug and device products, device and biological products, or biological and drug products;
3. A drug, device, or biological product packaged separately that according to its investigational plan or proposed labeling is intended for use only with an approved individually specified drug, device, or biological product where both are required to achieve the intended use, indication, or effect and, where upon approval of the proposed product the labeling of the approved product would need to be changed, for example, to reflect a change in intended use, dosage form, strength, route of administration, or significant change in dose; or
4. Any investigational drug, device, or biological product packaged separately that according to its proposed labeling is for use only with another individually specified investigational drug, device, or biological product where both are required to achieve the intended use, indication, or effect.

All three of these types of products, devices, drugs, and biologics, have different GMPs and regulatory requirements. So, if your company is developing a new medical product that will be regulated as a combination product, do you have to comply with all of the regulations pertaining to each type of product? Thankfully, the answer to this question is NO. Combination products are assigned to the FDA center that will have primary jurisdiction for its premarket review and regulation. The decision as to which Center will have primary jurisdiction for premarket review and postmarket regulation, or a lead center, is based on a determination of the "primary mode of action" (PMOA) of the combination product. The FDA defines PMOA as "the single mode of action of a combination product that provides the most important therapeutic action of the combination product." The PMOA also plays a role in deciding what GMP and quality system regulations will be required to be followed. In the case of a pharmaceutical company that wants to develop an auto injector system for one of their injectable drugs, the PMOA of the combination product is going to be the pharmacological action of the drug and the device (auto injector) is really just there as a delivery method. Since the company that is developing this combination product is a pharmaceutical company, they are already required to comply with the GMP for Finished Pharmaceuticals (21 CFR 211). However, now that they are manufacturing a combination product that includes a device they also have to concern themselves the quality regulations for medical devices that are set out in the quality system regulations (21 CFR 820). As we have pointed out on a number of occasions in this chapter, there is quite a bit of overlap between these GMPs and quality system regulations for different types of products and therefore, in the case of combination products, the FDA has identified the regulations that are specific to a type of product (in this case the medical device component) that the pharmaceutical must comply with in addition to the ones they are already required to comply with as a drug manufacturer. Therefore, in this case, where the PMOA has been determined to be the drug and the GMP and quality system being followed by the company is based on the manufacturing of the drug, the company must also comply with the following six provisions from medical device (21CFR Part 4).

Regulations (21 CFR 820) for their combination product.

1. 820.20: Management Responsibility
2. 820.30: Design Controls
3. 820.50: Purchasing Controls
4. 820.100: Corrective and Preventive Action
5. 820.170: Installation
6. 820.200: Servicing

In the case of a drug–device combination product, where the PMOA has been determined to be the device (i.e. drug eluting stent) and the GMP and quality system being followed by the company is based on the manufacturing of the device, then the company must also comply with the following eight (8) provisions from the GMP regulation for finished pharmaceuticals (21 CFR 211) for their combination product (21CFR Part 4).

1. 211.84: Testing and approval or rejection of components, drug product containers, and closures.
2. 211.103: Calculation of Yield.

3. 211.132: Tamper-evident packaging requirements for over-the-counter (OTC) human drug products.
4. 211.137: Expiration dating.
5. 211.165: Testing and release for distribution.
6. 6.211.166: Stability testing.
7. 211.167: Special testing requirements.
8. 211.170: Reserve samples.

12.5 CONCLUSION

Before any company starts a development program for a new drug, device, biologic, or combination product, it is important for them to understand what the regulatory agencies expectations will be with respect to ensuring the quality of their new healthcare product throughout the development cycle. One of the best ways to do this is to look at the GLP, GCP and GMP requirements for the specific type of medical product that you are developing.

The good practices (GxPs), when fully implemented, will employ a quality system of management controls for manufacturing facilities, research laboratories, clinical sites and the organizations that sponsor them, to ensure the uniformity, consistency, reliability, reproducibility, quality, and integrity of the data that is used to support the development, approval and marketing of these medical products.

12.6 USEFUL WEB SITES

FDA Guidance Documents	URL
The Applicability of Good Laboratory Practice in Premarket Device Submissions: Questions and Answers—Draft Guidance for Industry and Food and Drug Administration Staff	https://www.fda.gov/medicaldevices/deviceregulationandguidance/guidancedocuments/ucm366338.htm
1981 Questions & Answers—Good Laboratory Practice Regulations	https://www.fda.gov/ICECI/Inspections/NonclinicalLaboratoriesInspectedunderGoodLaboratoryPractices/ucm072738.htm
Guidance for Industry: Questions and Answers About the Petition Process	https://www.fda.gov/food/guidanceregulation/guidancedocumentsregulatoryinformation/ucm253328.htm
Good Laboratory Practice (GLP) Handbook	http://www.who.int/tdr/publications/documents/glp-handbook.pdf
Good Clinical Practices	http://www.ich.org/fileadmin/Public_Web_Site/ICH_Products/Guidelines/Efficacy/E6/E6_R2__Addendum_Step2.pdf
ISO 9001: Quality Management	https://www.iso.org/iso-9001-quality-management.html

Index

G

H

I

L

M

N

Printed in the United States
By Bookmasters